California Gold Milling Practices

By Edward Preston

with an introduction by Kerby Jackson

Introduction

It has been nearly one hundred and twenty years since the State of California released it's important publication "California Gold Mill Practices". First released in 1895 this important volume has now been out of print for over a century and has been unavailable to the mining community since those days, with the exception of expensive original collector's copies and poorly produced digital editions.

It has often been said that *"gold is where you find it"*, but even beginning prospectors understand that their chances for finding something of value in the earth or in the streams of the Golden West are dramatically increased by going back to those places where gold and other minerals were once mined by our forerunners. Despite this, much of the contemporary information on local mining history that is currently available is mostly a result of mere local folklore and persistent rumors of major strikes, the details and facts of which, have long been distorted. Long gone are the old timers and with them, the days of first hand knowledge of the mines of the area and how they operated. Also long gone are most of their notes, their assay reports, their mine maps and personal scrapbooks, along with most of the surveys and reports that were performed for them by private and government geologists. Even published books such as this one are often retired to the local landfill or backyard burn pile by the descendents of those old timers and disappear at an alarming rate. Despite the fact that we live in the so-called "Information Age" where information is supposedly only the push of a button on a keyboard away, true insight into mining properties remains illusive and hard to come by, even to those of us who seek out this sort of information as if our lives depend upon it. Without this type of information readily available to the average independent miner, there is little hope that our metal mining industry will ever recover.

This important volume and others like it, are being presented in their entirety again, in the hope that the average prospector will no longer stumble through the overgrown hills and the tailing strewn creeks without being well informed enough to have a chance to succeed at his ventures.

Kerby Jackson
Josephine County, Oregon
November 2014

3

NOTES ON GOLD MILLING IN CALIFORNIA.

The system of reduction of gold-bearing ores with stamps, as at present carried out in California, is the result of progressive improvement during the past forty-four years. The first successful mill in this State was built in the winter of 1850–51, and used steam for power.

Starting with the ancient Mexican arrastra, crushing, with the help of a mule and one man, a few hundred pounds of ore at a charge, we have progressed to the present aggregation of mechanical appliances, as seen in the modern stamp-mill, requiring great motive power, and disposing of hundreds of tons of ore in the course of a day. This progression is largely the result of accumulated practical experience on the part of the designers and builders of mills, as well as of the millmen in the handling of the various gold ores. Of late years scientific investigations have greatly aided in improving both the process and the mechanism.

That the results accomplished have been of economic value is evident from the fact that while formerly a yield of 30% to 40% of the total gold in the ore was the average obtained, the best mills of to-day are able to more than double these figures. That our methods may still be improved upon, and the margin of wasted gold be further narrowed down, is the point for which all intelligent millmen are striving. While the stamp-mill itself had been used for crushing ores long before the discovery of gold in California, since that time it has been greatly improved in detail, and its capacity and efficiency increased, hence what is now known as the "California gold mill" is a very different affair from the clumsy mills first used for crushing quartz in this State. The California gold-milling processes and the California millmen are, as a result, finding due recognition outside of their own immediate field of operation, as is evidenced by the increasing outside and foreign demand for our men and milling machinery. The development of the milling process, keeping pace with the improvement of the machinery required for ore reduction, has had the beneficial effect of greatly lessening the working expenses, permitting ores of a low grade to be worked at a profit. California has a great abundance of this class of ores, comparatively untouched, and these must be mainly relied on in the future as the sources of the precious metal. Already, under extremely favorable conditions, ores are being mined and milled in California at a cost of 50 cents per ton, as at the Spanish Mine, in Nevada County, where, with Huntington roller mills, ores yielding 85 cents per ton have been worked at a profit.

MILL SITE.

When assured of a constant and sufficient supply of ore, it is of the greatest importance that the site for the mill should be chosen with due regard for economic treatment. This necessitates the observance of the following points: The means of transportation of the ore from the mine to the mill, which should be done automatically, or at least with as little handling as possible, conveying the ore at once to the highest point in the mill, so that it will descend by gravity from one to the other in all the different consecutive operations. Another important feature is to provide sufficient space for capacious ore-bins, which are necessary to prevent a stoppage of the mill through a lack of ore, caused through unavoidable delays in the mine or along the roads. The accessibility of the mill site as regards fuel, water, or electrical transmission, according to the motive power to be used, and their continuity and cost at all seasons of the year, must likewise be considered. The possibility of placing the levels for the different floors on solid rock foundations should be investigated, as stability of the machinery is most essential for successful milling.

The ideal site would be to have the mill in close proximity to, but below the level of, the collar of the shaft or the mouth of the tunnel, on sloping ground, where the ore can be delivered directly from the mine to a " grizzly " on the upper floor of the mill, to be passed later, without rehandling, through the crushers, ore-bins, self-feeders, mortars, etc., while leaving sufficient space for a waste dump. For a mill arranged in this manner, including concentrators and canvas platforms, 40' of fall should be available. If chlorination works are also to be used, a greater fall is desirable.

MILL CONSTRUCTION.

After deciding on a suitable site, the surface should be removed down to the bedrock and leveled off for the different floors. Solidity and accessibility are the chief points to be observed in placing the different parts of the mill. Where required, heavy stone walls should be erected as buttresses. The foundation for the mortars and the proper erection of the battery frames are points requiring particular attention. For the mortar-block, a trench is prepared of suitable depth, preferably in solid bedrock, proportioned to the height of the block, and wide enough to leave about 2' of free space around it, which is later filled in with concrete or tailings from the battery. These mortar-blocks vary from 8' to 15' in length and are dressed at the upper end to the size of the bed-plate of the mortar. In California they can be obtained frequently from a solid cut of a pine tree, or else consist of two or three sawed blocks fitted and bolted together; but where clear timber of the requisite size is difficult to obtain, the block can be constructed of 2" plank, as is done in the Black Hills in Dakota.* There the bottom of the trench for the block is leveled and some sand tamped down, on which two layers of 2" plank are placed crosswise and spiked to each other, and made perfectly horizontal. On this foundation a mortar-block is constructed of 2" planks, from 11' to 14' long, according to the depth of the trench. The planks, which should be of clear lumber, and varying breadths (in order to break joints), stand on end, with their width parallel to the long side of the mortar. They are spiked together and fastened above and below with binders bolted to each other by transverse rods; the upper binders (8" x 12") being even with the top of the mortar-blocks; the lower binders (12" x 12") are 3' lower.

The top of the mortar-block should be planed perfectly true and

*Gold Milling in the Black Hills, by H. O. Hoffman. Transactions of the American Institute of Mining Engineers, Vol. 17, 1888–89.

leveled, and where several blocks are placed in line all the blocks should be sawed off to one height. Before setting the mortar upon the block a sheet of rubber cloth, $\frac{1}{4}''$ thick, should be placed between, or when this is not obtainable, two or three folds of mill blankets, well tarred, will answer the purpose.

The mudsills should be of square timber, free from sap, bedded in concrete on the bedrock and secured by anchor bolts to the foundation; they also should be bolted to the linesills.

The uprights of the battery frames are supported in various styles: with diagonal braces and hog chains at front or back, or with so-called knee-frames. In the former style the brace is placed on the same side as the counter-shaft, which rests low down on the battery-sills. This style is well suited for small mills using stamps not to exceed 750 lbs., but for large mills using heavy stamps the knee-frames are the more suitable, with the counter-shaft on a level with the cam-shaft. What is known as the reversed knee-frame forms a strong, compact construction, but requires the counter-shaft to rest on the battery-sills behind the frame. The accompanying cuts show the construction of back and front knee-braces as supplied with mills from the Union Iron Works of San Francisco.

Fig. 2 presents a view of a back-knee battery frame with a Union mortar and square ore-bins, showing the latest arrangement of working the self-feeder from a collar on the stem, instead of having the tappet strike the bumper.

Fig. 3 shows the back-knee battery frame with the cams revolving toward the ore-bin, with a Hayward mortar.

Fig. 4 shows the suspended ore-feeder with back-knee frame.

Fig. 5 is the arrangement of a front-knee battery frame, with Union mortar.

Figs. 6 and 7 show the method of securing knees to the battery posts.

The battery posts are made 24" deep, and from 12" to 20" wide; the center one of a ten-stamp mill being made the heaviest, as having to bear the greatest strain. They are let into the sills and secured to the line timbers by bolts. Besides the braces, the posts are given stability above the mortar by the guide-timbers (see Fig. 6), which extend from end to end in one piece, and are let into the posts to which they are bolted. The lower one is placed about 6" above the upper edge of the mortar, and the center of the upper one is about 3' from the top of the post. The seat for the cam-shaft bearings is cut in the upper part of the posts.

After lowering the mortar on the block, with the planed bottom resting evenly on the sheet of rubber cloth or folds of tarred blanket, it is fastened perfectly rigid by eight bolts, four on each long side, passing through the flange, which is cast on the bottom of the mortar. This flange is 4" wide and about $2\frac{1}{4}''$ thick. The feed floor should be high enough so that these bolts can be conveniently reached, to permit their tightening when required. The journals for the cam-shaft, which are placed in the recesses cut out of the battery posts for their reception, are lined up and "babbitted" prior to receiving the cam-shaft with its cams. The stems are placed from $\frac{1}{2}''$ to 1" from the cam-shaft, and just far enough from the cams to clear them when dropping. The cam-shaft is made of wrought iron or soft steel, from $4\frac{1}{2}''$ to 5" in diameter, turned true, and should have key-seats for securing the cams. There should

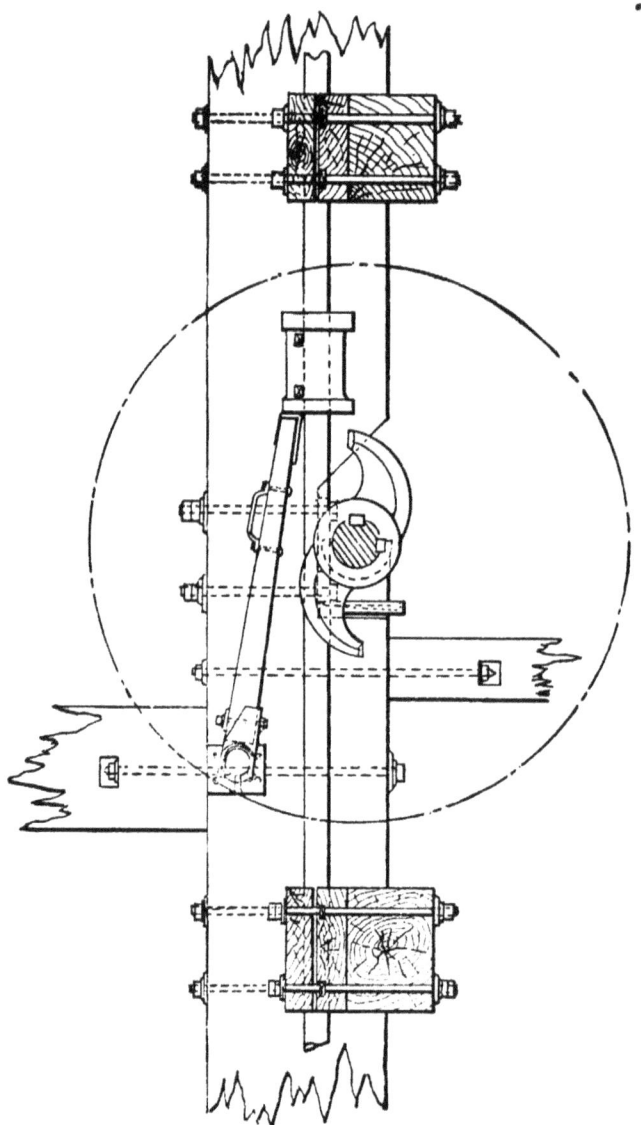

FIG. 6. METHOD OF SECURING KNEES TO BATTERY POSTS.

be two key-seats, and placed one third of the shaft circumference apart. At one end of the cam-shaft the cast-iron "hub" of the belt pulley (with flanges) is keyed on. This pulley is built of wood, and turned true on the shaft. Where there is more than one battery to the mill, it is best to have a cam-shaft for each ten stamps, as this permits of repairs, such as changing cams, etc., without stopping more than ten stamps.

The guides (see Figs. 6 and 7, and D, E, and F of Fig. 8), which direct the drop of the stems, are in two sets, upper and lower—the former above the tappets, and the latter below the cams—and are bolted to the guide-girts by eight bolts. They are best made of hard wood, but pine answers sufficiently well—though the former lasts five times as long as pine. The old style guide (F, of Fig. 8) consists of two pieces of 4" plank 14" wide, planed on all sides, and of sufficient length to fit easily between the battery posts, with equi-distant semi-circular grooves, fitting together, for the passage of the stems. A quick and exact way to make these grooves is to clamp the two planed planks tightly together as they are to be placed on the frame, set them on edge, and, after marking off the centers for the five stems, bore out the circle

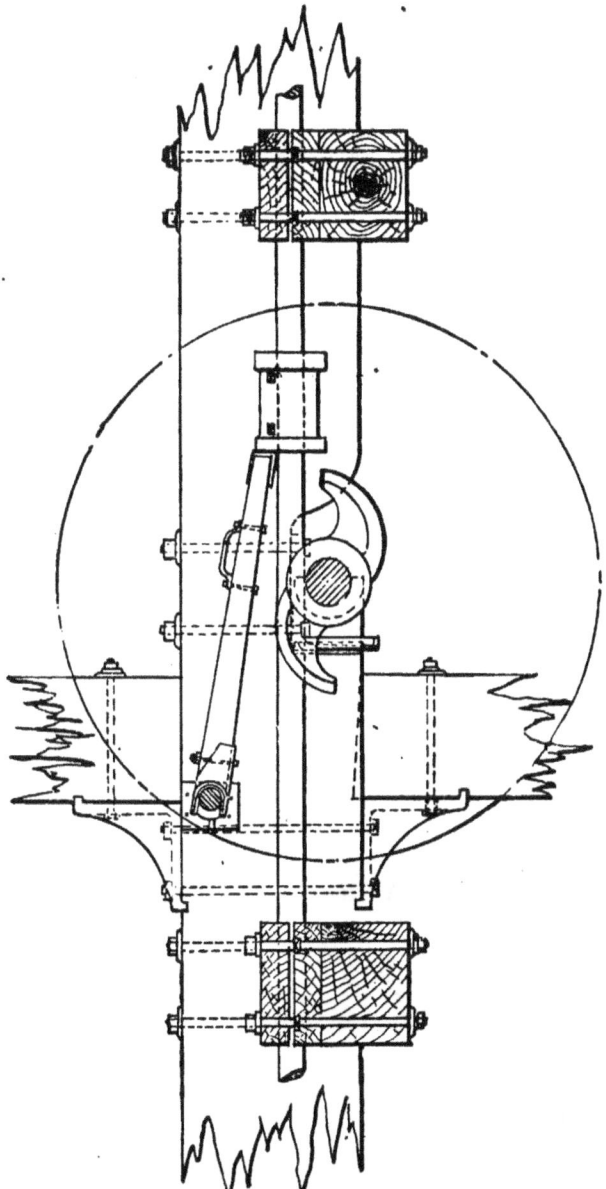

FIG. 7. METHOD OF SECURING KNEES TO BATTERY POSTS.

(using the joint line for the point of center) with a long-handled auger having an adjustable bit.   These are kept in many mills for this special purpose.

Before bolting the guides in place, half-inch pieces are placed between the two halves, and adjoining each stem, which are planed down later as the guides wear, leaving but little play for the stem.   After boring out the grooves for the stems, and before putting the guides in place, they should be lubricated.   A convenient and economical plan is to cut some semi-circular pieces of thin sheet iron of a somewhat larger diameter than the grooves, and drive them into the wood at both ends of the channel; then lay the halves level, groove side up, and fill the latter with linseed oil, letting them remain until the wood has taken up all it will absorb, when the remainder is returned to the can, and the sheet-iron pieces removed.

If, in a pine guide, those portions occupied by the grooves are cut out square, and hard-wood bushings fitted in before boring out, the stem will work parallel to the wood fiber, which reduces the friction,

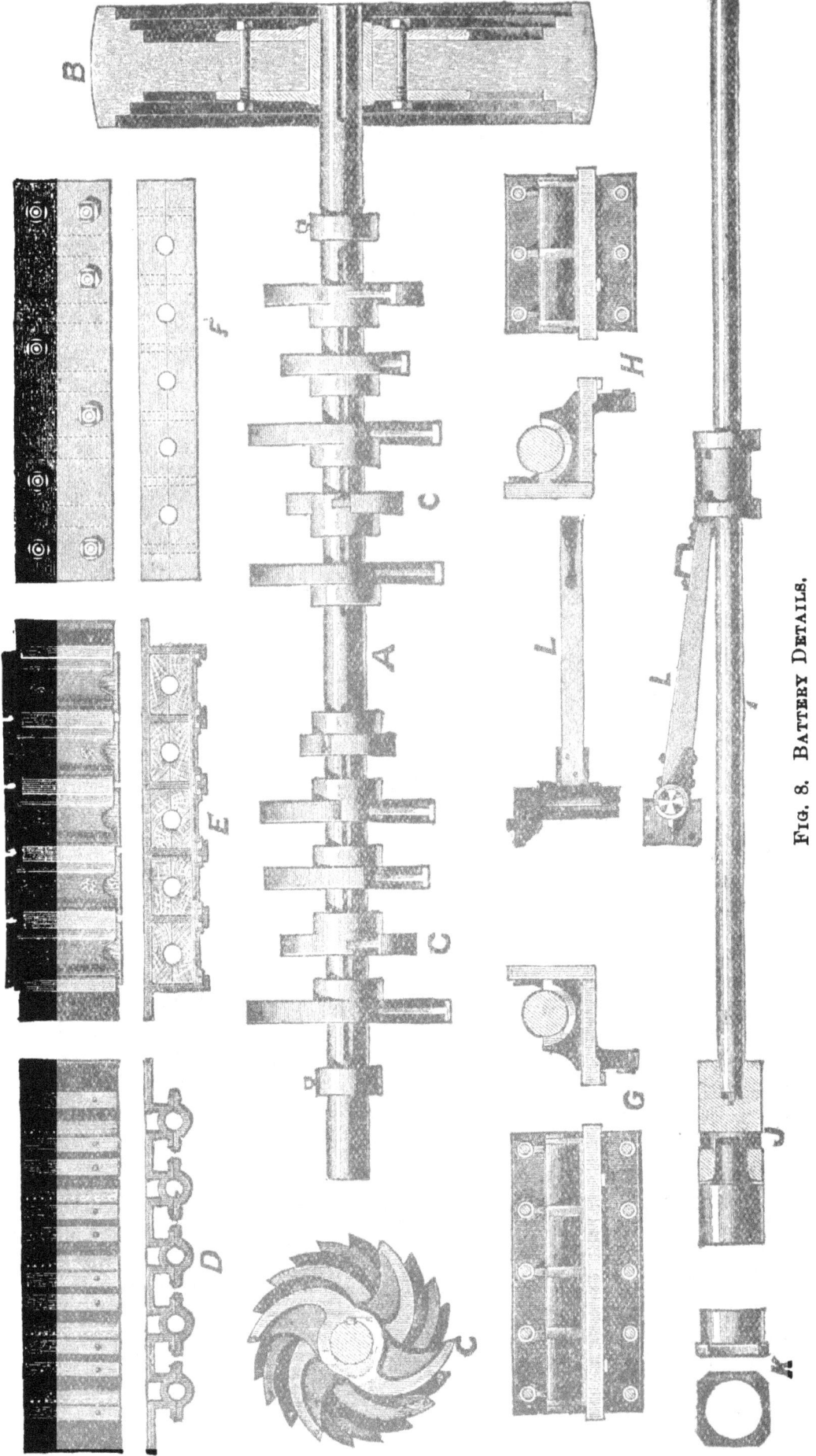

Fig. 8. Battery Details.

and lengthens the life of the guides; while only the hard-wood bushings would need replacing, making the cost of the guides less.

The great drawback to these guides is that when a stem has to be removed, the entire battery has to stop; hence the adoption of separate guides for each stem, being either all iron (see D, Fig. 8), or wooden bushing in iron frames (see E, Fig. 8), which are held in place by wedges and the lips of the iron frames.

For the support of the stamp-stems when suspended, wooden latch-fingers, or jacks, are supplied. (See Figs. 6 and 7, and L, of Fig. 8.)

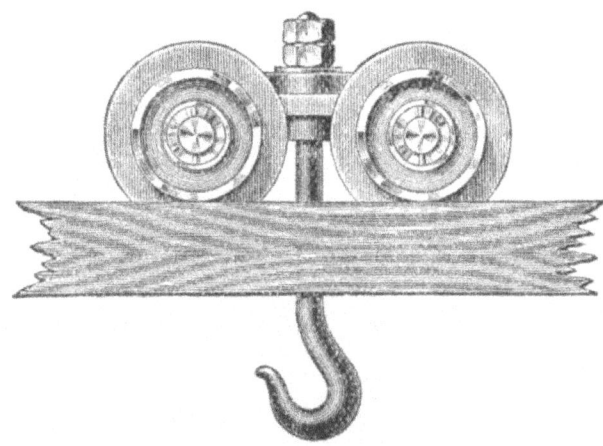

FIG. 9. OVERHEAD TRAVELING CRABS.

A jack-shaft, 3″ in diameter, rests in bearings attached to the inner sides of the battery posts; on this, cup-shaped sockets ride, in which the wooden fingers are attached, shod at the upper end with an iron plate ⅛″ thick, and provided with an iron or leather "hand-hold" near the top.

For the greater convenience of quickly removing and replacing stems, or cam-shafts, large mills are supplied with overhead travelers, or "crabs," in line with the batteries, in connection with a chain block and tackle running on plates secured to the roof, shod with iron tracking. To easily reach the cams, tappets, etc., a platform is placed just below the cam-shaft.

The feed-floor consists of a double board floor of 1″ lumber, with broken joints, supported on joists 18″ apart, and about 2′ below the feed opening in the mortar.

MILL DETAILS.

The Grizzly is a coarse screen consisting of a number of parallel bars attached to a frame, set on an angle from 45° to 55°, over the ore-bin. These bars may be of round, rectangular, or V-shaped (apex down) iron, or of wood, faced with iron, and resting on several iron cross-rods, held apart with iron washers; the distance between the bars should be equal to the opening the rock-crusher jaws are set to—from 2″ to 3″. There are no fixed dimensions of length or breadth, as these depend in a measure on local conditions; but they are usually from 3′ to 6′ wide, and long enough (12′ to 15′) to give the fine material time to drop through the spaces before reaching the crusher floor.

Where substantial steel T rails are used for tracking in the mine, they can be made to serve for grizzly-bars when no longer of use in the mine, by turning them with the base up.

The grizzly should be placed at the highest point of the mill over the ore-bin, where the car or wagon can enter and dump. Its chief object is to separate at once the finely divided ore from the coarser; a secondary purpose is served in affording an opportunity to recover drills, gads, or hammers that may have come from the mine, in the ore, before they reach the rock-breaker or mortar. Its lower end rests on a platform in front of the rock-crusher, or better, in a chute with an adjustable end-gate placed above the mouth of the rock-crusher so as to permit of its being fed automatically.

Where the ore as delivered from the mine carries less than 5% of fine stuff, the grizzly should be dispensed with, especially where, in con-

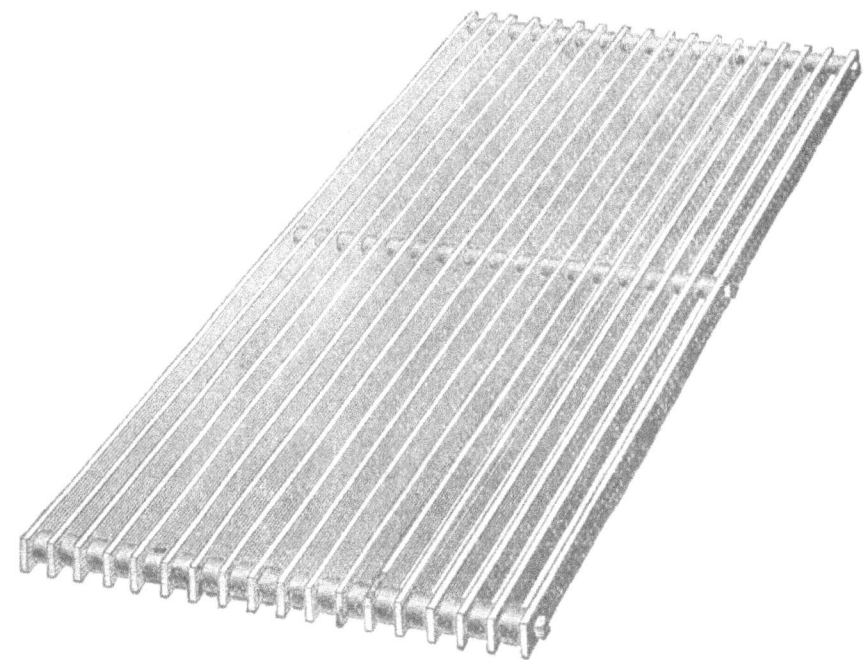

FIG. 10. GRIZZLIES.

structing the mill, fall must be economized. Some object to the use of the grizzly, as tending to feed all the hard rock by itself, and say the output of those batteries is below the others.

Rock-Breakers or Crushers are placed on a platform below the grizzly and above the ore-bin in such a manner that the crushed rock mingles with the fine stuff passing between the bars of the grizzly. The rock-breaker must be of sufficient weight to remain firm in its place, and strong enough to resist heavy strains; the dies should be easy to exchange and adjust, and all parts requiring to be oiled should be arranged to prevent oil coming in contact with the quartz. In large mills it is best to have one crusher to supply every twenty stamps, and on account of their intermittent work, they should have driving power separate from that of the stamps.

Rock-breakers are adjusted to crush the rock smaller than the throat of the mortar (therefore, less than 3″), but as the work of the rock-breaker is cheaper than that of the stamp, it would pay, with very hard rock, to do more of the crushing with this machine, even to the extent of placing two crushers, one beneath the other, and bringing the quartz greatly reduced to the stamps.

There are two general types of rock-crushers. The older pattern carries a flat, fixed jaw, working with one having a reciprocating motion and using flat or corrugated dies that are reversible. The Blake is representative of this pattern. The other pattern has an outer, circular, fixed jaw, within which a corrugated jaw circles, of which the Gates is representative. This latter machine permits of larger blocks being fed. It is an excellent machine for heavy work, and where the rock is not wet or clayey; but it requires greater horse-power, for where a Blake, 10″ by 8″, crushing 3 tons per hour, requires 9 H. P., the Gates, with a diameter of 37½″, crushing 3½ tons per hour, requires 16 H. P. The Gates consists of a nearly vertical shaft of forged steel, rotated from below by a

FIG. 14. ORE-BIN GATE.

beveled wheel set ½″ out of center, on the top of which a chilled-iron conical head is attached, with the base downward, rotating within chilled-iron concaves, with an outward slope, set in the cylindrical body of the machine. Between these two faces the ore is crushed, their distance apart below being gauged by set-screws. The shaft, by being made to revolve around an eccentric at the bottom, has a constant crushing power without doing any grinding. A set of concaves lasts two years, and can be replaced; the center shaft with the chilled-iron head has been known to crush 120,000 tons of an average hard quartz before wearing out.

Ore-bins should always be as spacious as the surroundings will permit, but never of less capacity than will carry a twenty-four hours supply for the mill, say about 65 cu. ft. to the stamp. They are usually constructed with a sloping bottom, to facilitate discharging, but where very large bins can be erected this feature is not so essential. These bottoms must be solidly braced and ought to be covered with iron plates over those portions where the ore has to be dropped. The front of the bin is parallel with the mortars and supplied with gates for each battery above the level of the hopper of the self-feeders. These gates should be regulated by a pinion and rack, and set for a regular discharge and delivery, through chutes, into the self-feeders. The chutes should be lined with heavy sheet iron.

FIG. 15. HENDY CHALLENGE ORE-FEEDER.

Self-Feeders.—The entire value of the stamp battery hinges on a regular and even feeding, and as it can be done much better (from 15% to 20%) by a machine than by hand, this latter method has become well nigh extinct in California.· Among the mechanical feeders mostly used are the Challenge (in two patterns), Tulloch, Stanford, and Roller feeders.   Although the three latter are very serviceable for certain classes of ore, and are cheaper in first cost, the Challenge is undoubtedly the best all-round machine, which is proved by its almost universal adoption.   They are either placed on a frame which runs on an iron track in the feed floor, back of and at right angles to the battery, or are suspended from tracks supported by the battery posts and standards placed against the ore-bin. This latter pattern permits of greater accessibility to the feed side of the mortar.   In general, the Challenge feeders consist of a hopper with a movable circular plate beneath, set slightly inclined toward the mortar, receiving a rotary motion by means of gear wheels acting on the lower face of the plate, which are moved

FIG. 16. TULLOCH AUTOMATIC ORE-FEEDER.

by a friction grip that receives its impetus from a blow of the descending stem on a bumper-rod connected with it. Movable wings extending from the point of the hopper over the plate toward the throat of the mortar permit a given quantity of the ore to be scraped off at each blow through a partial rotation of the plate. The older machines were made with right or left-handed bumpers, but the present and better plan is to place the rod in the center, so that the third stamp in a five-stamp battery imparts the blow. The newest machines have no bumper-rod, but are worked by a collar fastened on the stem above the top of the mortar. Each battery is supplied with its own self-feeder.

The Tulloch feeder consists of a square frame, into which a hopper fits, having below a tray suspended from the frame at any desired angle, and in such a manner as to have a forward and backward swinging motion inside the frame, which can be arrested on the forward motion at a certain point by lugs, underneath the tray, striking a bar. The back of the hopper is supplied with an adjustable scraper, and at each motion of the tray a certain amount of the ore is scraped forward and falls into the battery. The machine is operated by the descent of the stamp.

Mortars.—The mortars in California are mostly single-discharge, and cast in one piece, extremely solid. When required in places inaccessible by wagon roads, they are cast in pieces, which are later bolted together. Their interior form depends on the nature of the ore, and the procedure to be applied; thus we find them made with narrow or flaring, deep or

2—GMP

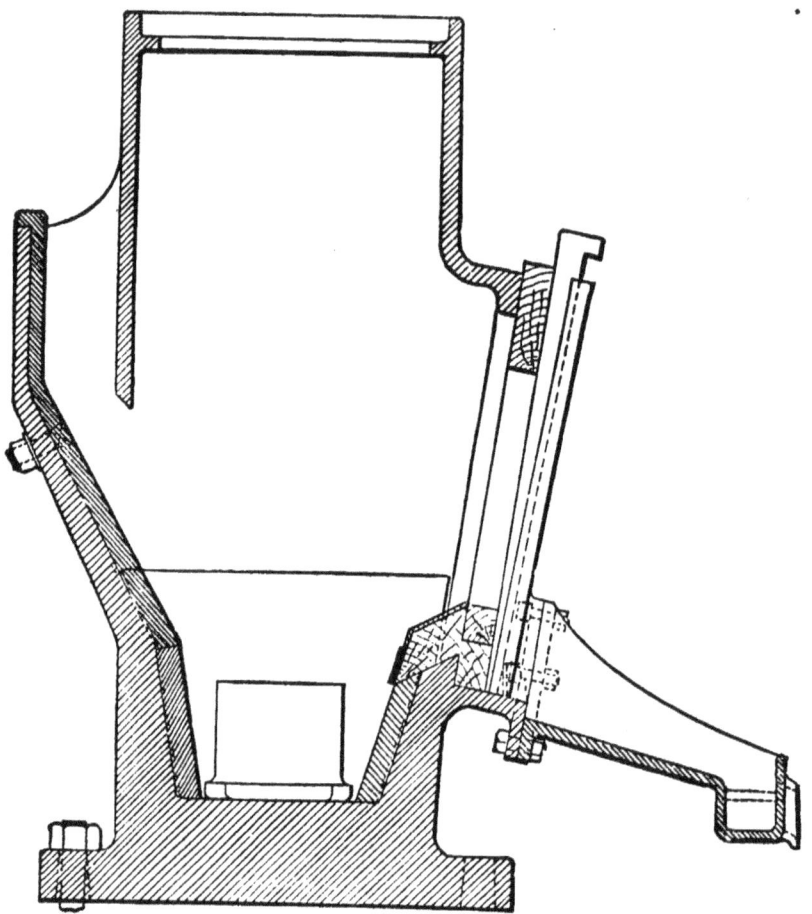

FIG. 17. HAYWARD MORTAR.

shallow troughs, and with or without inside plates. Mortars with narrow troughs are made for greater output, while a wide trough assists battery amalgamation, and gives opportunity for placing inside copper plates. In some of the newest styles of mortars a series of grooves are furnished in the lining plates, to contain quicksilver. The mortars weigh from 4,000 to 6,500 lbs., the bottoms being made extra heavy; in some of the latest patterns the bottoms are 8″ thick. The length varies between 4½′ and 5′, and the height from 4¼′ to 4½′. The inside width of the trough corresponds with that of the foot-plate of the dies. A heavy flange, 4″ x 3″, is cast on the base of the long sides, in which are four holes on each side for bolts, to secure the mortar to the block.

The difference in design hinges chiefly on the different opinions of leading mining men as to the method and value of amalgamating inside the battery.

Figure 17, known as the Hayward mortar, is a full-lined mortar, with flaring trough, weighing, complete, about 6,500 lbs., without any special arrangements for inside amalgamation.

Figure 18, the Alaska mortar, is a full-lined mortar, with flaring trough, in which the linings are furnished with grooves, to contain quicksilver.

Figure 19, the Wilman's mortar, was the first attempt at inside amalgamation, using an inside removable copper plate, but this failed to work well; the copper plate, being so close to the shoe and die, scoured, and could not retain the amalgam. In remedying this defect, the

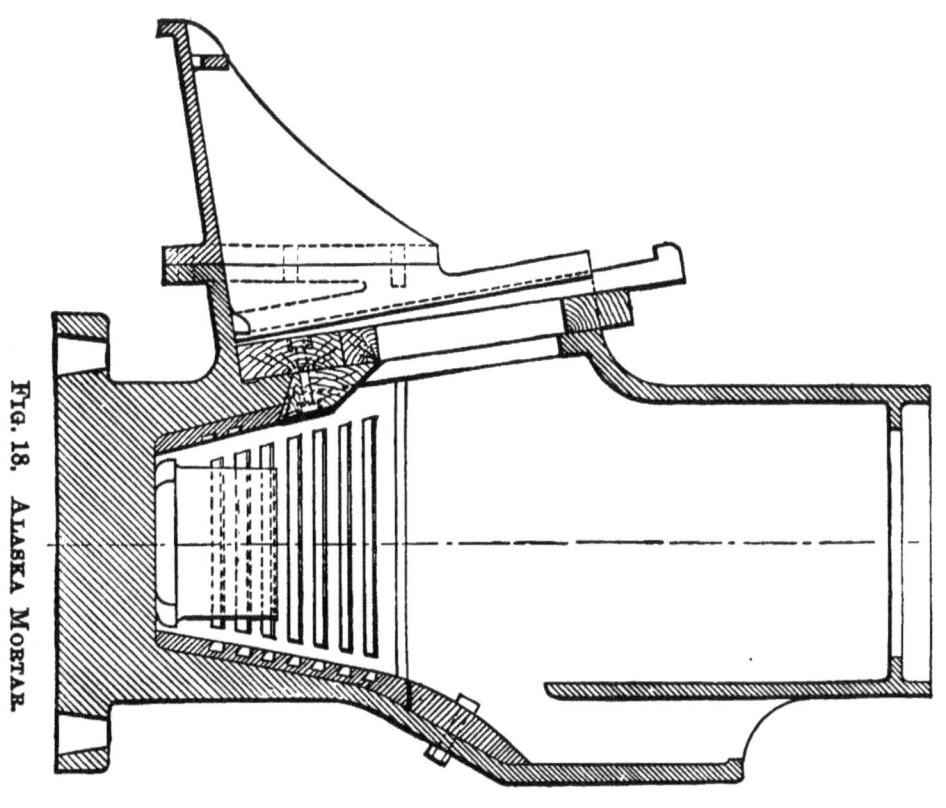

FIG. 18. ALASKA MORTAR.

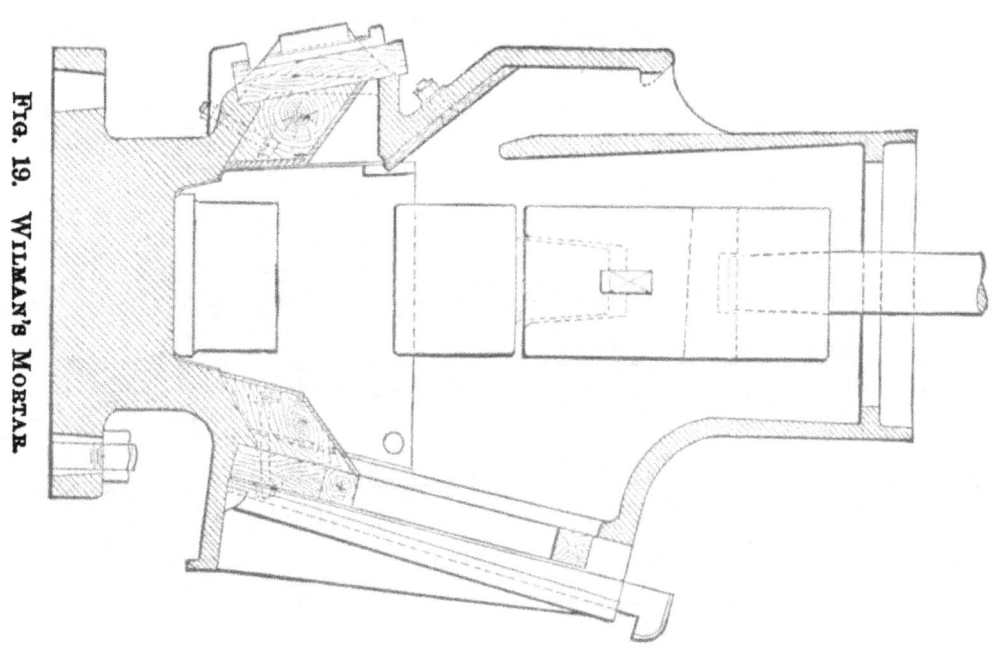

FIG. 19. WILMAN'S MORTAR.

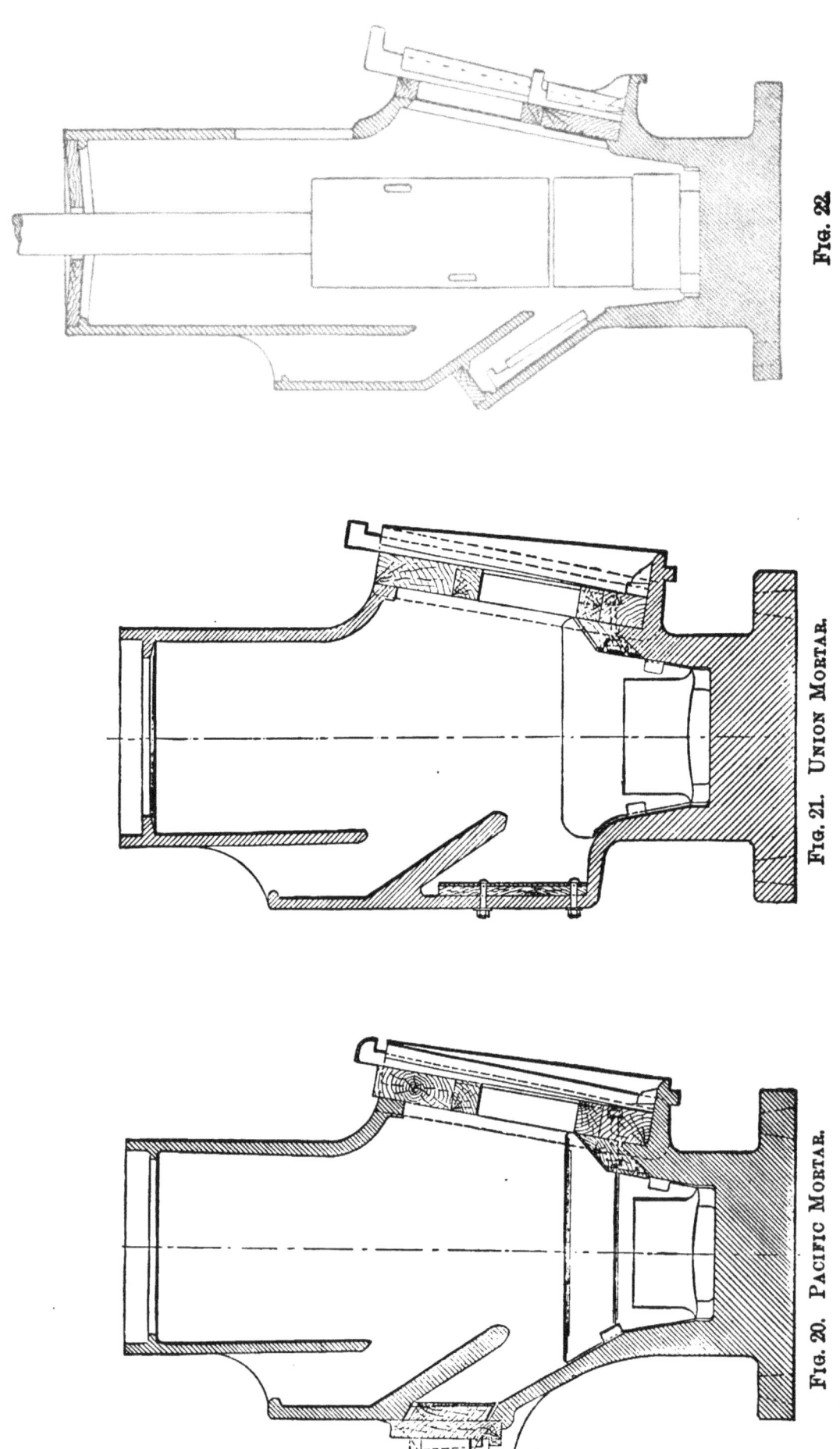

FIG. 22.

FIG. 21. UNION MORTAR.

FIG. 20. PACIFIC MORTAR.

Pacific mortar (Fig. 20) was produced, in which the copper plate was placed higher up, under the feed throat; experience in the working of which, suggested changes that finally evolved the Union mortar (Fig. 21), which is provided with a copper amalgamating plate, 12"x48", bolted in a recess at the back of the mortar, reaching below the level of the screen opening. This plate is removed on clean-up days to be scraped, and then replaced. Similar good results have been obtained by the writer in using a double-discharge mortar, and filling the back discharge opening with a plank instead of a screen, to which a plate, 8" wide and the length of the opening, had been attached.

A modification of this back-plate arrangement is shown in the accompanying drawing (Fig. 22) of a mortar designed in Milwaukee, Wis. Here the back plate is accessible, from the back of the mortar, through a covered opening; it is secured in place by a dovetailed key at each end, allowing it to be adjusted to the varying height of the dies.

Dies (see K, Fig. 8, and S, Fig. 23).—They consist of a cylindrical body of the same diameter as the shoe, with a square foot plate with broken corners, and should fit loosely against the front and back plates of the mortar. The broken corners permit their easy removal. They are cast both in iron and steel.

Shoes (see J, Fig. 8, and P, Fig. 23).—They are made of iron or steel, and consist of a cylindrical body of the same diameter as the stamp-head, with a cone-shaped neck, half as wide as the cylindrical body, and about 5" long. The weight of the shoe bears a certain relation to the other parts of the stamp, generally about one sixth of the total weight when made of chrome steel, but somewhat less when made of iron. The cylindrical portion of the shoe is somewhat longer than the corresponding part of the die, on account of its greater wear—the latter being protected by a cushion of quartz. Both shoe and die are used until worn as thin as possible; with the shoe, this may be $\frac{1}{2}$", though rarely, while the die is worn to the foot-plate, if not fractured previously. This practice is not to be commended, and should only occur in case of necessity. On stamps weighing about 900 lbs. the shoes, if of chrome steel, weigh about 150 lbs., and if of iron, weigh about 20 lbs. less, and are about 9" in diameter. The life of the shoes depends on the nature of the quartz and the height and speed of the drop, but as a general rule shoes and dies of steel last as long as two and a half sets of iron ones, and cost twice as much. In the matter of choice between steel and iron, the vicinity of the mill to foundries is of consequence. Steel shoes and good iron dies usually work very smooth, but where the waste iron can be disposed of at a foundry, this metal is preferred for both.

Stamp-Heads, Bosses, or Sockets (see J, Fig. 8, and R, Fig. 23).—They are made of cast iron or steel, of the same diameter as the cylindrical part of the shoes and dies, with two conical sockets; the upper one accurately bored out to contain the tapering end of the stem, and the lower one to receive the neck or shank of the shoe with its inclosing circle of thin wooden wedges. Transverse, rectangular keyways, at right angles to each other, pass through the stamp-head at the end of the conical openings, connecting therewith in such a manner that when both stem and shoe are attached to the boss they protrude into the keyways. This enables them to be forced out by the driving in of a wedge-shaped

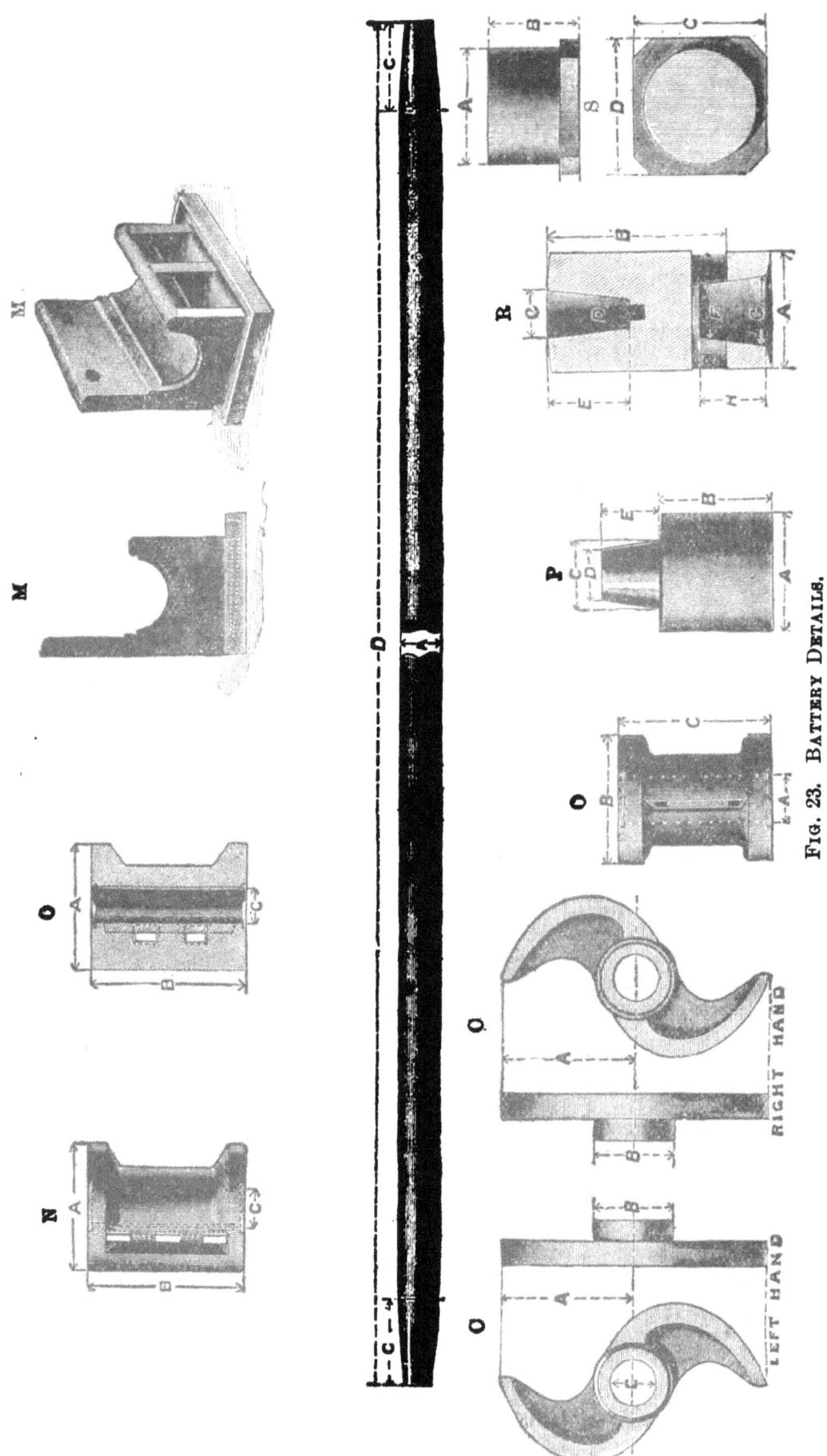

FIG. 23. BATTERY DETAILS.

steel drift about 1″ wide, 18″ long, and tapering down from 2″ toward the point. The ends of the stamp-head are usually reinforced by having iron bands shrunk onto them.

Stems (see I, Fig. 8, and D, Fig. 28).—They are made of wrought iron or soft steel, turned perfectly true, and tapered at both ends for a distance of 6″ or 8″. They are from 11′ to 14′ in length, the diameter varying with the weight of the stamp from $2\frac{7}{8}$″ to $3\frac{1}{2}$″. They are reversible, so that if one end breaks, the other end can be used before sending to the shop for repairs. When this repair is made, the whole stem should be annealed. The stem carries the greater weight of any part of the stamp, amounting to nearly one half. The stems hang in the guides at even distances from center to center, and are supported while at rest by props or fingers catching on the under face of the tappets.

Tappets (see O, N, and O, Fig. 23).—They are made of tough iron or steel, cylindrical, with a flange on both ends, and accurately bored through the center, a shade wider than the stem, and counter-bored at both ends; they are provided with a rectangular recess adjoining the central bored hole, 7″ to 8″ long, and from 2″ to $2\frac{1}{2}$″ wide, in which a gib is fitted. This is a piece of wrought iron or steel, grooved on one side, with curvature $\frac{1}{8}$″ smaller than that of the stem, and planed flat on the opposite side. Two, or in some cases three, slots are cut through the tappet between the flanges, at right angles to the stem, which connect with the rectangular recess for the gib, so that keys, when driven through the slot, press the gib against the stem, which should slide smoothly through the center of the tappet. The tappets are faced on both ends, and are reversible. The keys are of steel, fitted and marked. Tappets weigh from 100 to 120 lbs. When fastening the tappet, the keys are driven in solid; but care must be observed, as when too tightly keyed, the tappet is liable to split.

Cams (see C, C, Figs. 8 and 23).—They are of tough cast iron or steel, double armed, and strengthened by a hub; which latter is frequently reinforced by having a wrought-iron ring shrunk on. The cam itself is the involute of a circle having for its radius the distance between the center of the stem and the cam-shaft, somewhat flattened, however, at the point of the cam. It comprises a face from 2″ to 3″ wide, ground off, decreasing in thickness from the hub to the point, and strengthened by a rib on the under side, which runs from a point to several inches deep at the hub. The cam is fastened to the cam-shaft by steel, hand-fitted keys.

Cam-Shafts (see A, Fig. 8).—They are made of wrought iron or soft steel, turned true, with double key-seats, 120° apart for cams, besides key-seats for the driving pulley. The cams are slipped on the cam-shaft with the hub side away from the stem, and keyed solidly in their respective places; they must be placed in such a manner that when the cams are raising the stamps, the weight is as nearly evenly distributed over the shaft as possible. For this reason proper attention must be given to the sequence in which the stamps are to drop in the battery. Where the shaft is for ten cams, the following order or succession of drops is recommended, viz.: 1, 5, 9, 3, 7—10, 6, 2, 8, 4, and would give a drop in each battery as follows: 2, 4, 1, 3, 5.

The cam, in picking up a stem on the under side of the tappet, imparts a revolving motion to tappet and stem, requiring from four to six strokes of the cam to complete one entire revolution. A too rapid revolution indicates the need of lubricating. The revolving of the stem assists in giving an even wear to the faces of the shoes and dies, but it does not impart a grinding action to the stamp, as frequently stated, which can be proved by holding a piece of chalk against the stem during its ascent and descent.

Screens and Frames.—Screens of different materials and with different orifices are used; the materials comprise wire cloth of brass or steel, tough Russian sheet iron, English tinned plate, and, quite recently, aluminum bronze. The Russian sheet-iron plates are perforated with round holes or slots; the latter are vertical, horizontal, diagonal, or curved, and are either entirely smooth or burred on the inner side. The latter form is intended for longer wear by closing the burrs with a mallet when too large, thereby prolonging the life of the screen. These screens last from fifteen to thirty-five days. The plates have glossy, planished surfaces, and come in sheets of 28″ to 56″, costing in San Francisco from 65 to 80 cents per square foot. The English tinned-plate screens come in sheets of 1′ to 1½′ square; they are more flexible than the Russian iron, hence do not permit of the pulp caking along the lower edge when fed high; and, as compared with a Russian iron one of the same perforations, they give a greater discharge, but they are short lived—averaging about ten days. The tin is burned off before using. Brass screens, costing in San Francisco 36 cents a square foot, are sold in rolls; they give the greatest discharge for an equal area, and last from ten days to two weeks, but should not be used if cyanide of potassium be used in the battery, on account of clogging with amalgam. The "aluminum bronze" plates come in sizes similar to the sheet tinned plate, but unpunched, the latter work being done here; they are much longer lived than either of the other kinds, and have the further advantage that when worn out they can be sold for the value of the metal for remelting; these plates are bought and sold by the pound, and are said to contain 95% of copper and 5% of aluminum. Steel-wire screens are not much used, on account of their liability to rust. The life of a screen depends, aside from the manner of feeding, on the width of the mortar, the height of the discharge, and the hardness of the rock. Wide mortar and high discharge are favorable to the preservation of a screen; the form of the perforations—round holes, or slots, etc.—influences the discharge area of the screen.

A good deal of confusion exists in interpreting the numbers of the different kind of screens. Wire screens take their numbers from the meshes to the linear inch, while perforated and slotted screens are numbered from the needle used in punching them, these needle numbers being the same as are used for sewing-machines. The sizes most frequently used in gold milling are from No. 6 to No. 9 of the perforated and slotted screens, and from No. 30 to No. 40 of the wire screens. The slots are from ¼″ to ½″ long, and placed alternating or even in the rows, some being burred on the inner side.

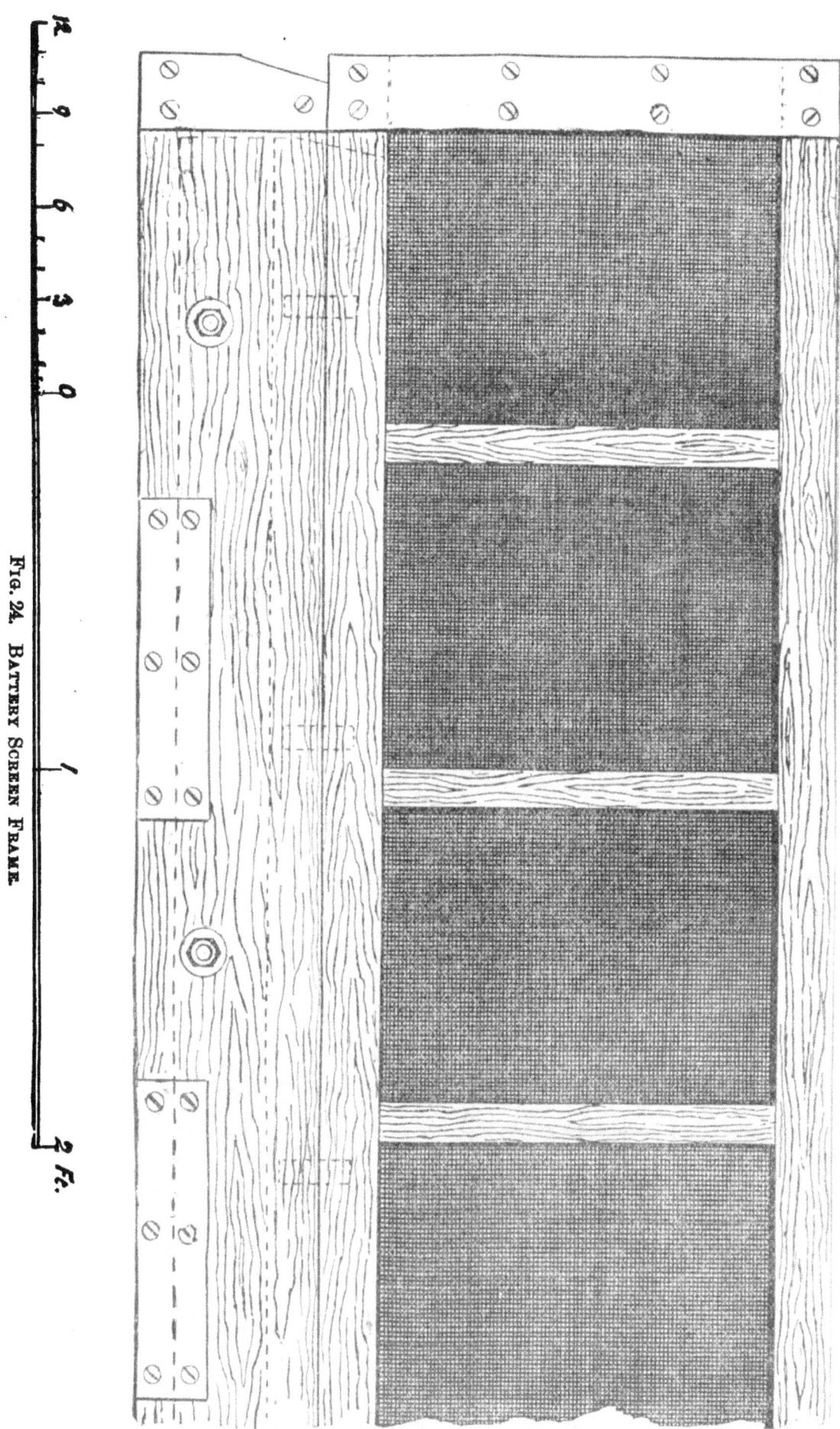

FIG. 24. BATTERY SCREEN FRAME.

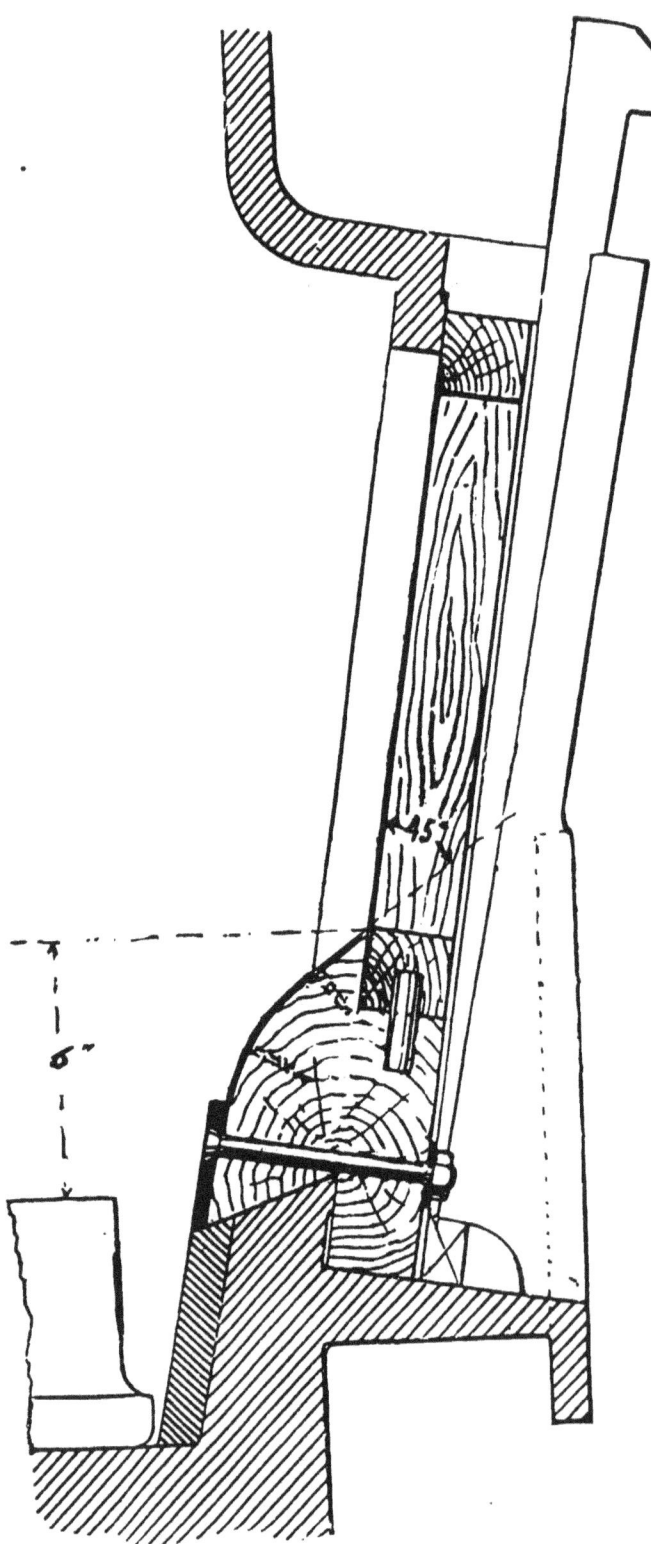

FIG. 25. ADJUSTABLE BATTERY SCREEN.

As dies wear down, wooden chock-blocks (on which the inside plates are fixed), of less
height, are substituted, thereby preserving uniformity in height of discharge.

The following table gives a comparison of the different varieties, with their numbers:

No. of Needle.	Corresponding Mesh.	Width of Slot. (Inches.)	Weight per Square Foot.
5	20	0.029	1.15 lbs.
6	25	0.027	1.08 lbs.
7	30	0.024	0.987 lbs.
8	35	0.022	0.918 lbs.
9	40	0.020	0.827 lbs.
10	50	0.018	0.735 lbs.
11	55	0.016	0.666 lbs.
12	60	0.015	0.666 lbs.

The proper size required is a matter for the millman to decide at each mill. The character of the ore and the coarseness of the gold have to be considered, as well as the inside dimensions of the mortar; ore carrying extremely fine gold requiring a finer crushing, as the gold must be freed from the quartz matrix in part if the quicksilver is to act on it; but where this would lead to if carried out to its legitimate end may be imagined, when the writer states that he has observed, under the microscope, a particle of quartz that had passed through a No. 9 screen (40-mesh) and still contained several separate but included particles of gold. Sulphide ores, having a much greater tendency to form slimes, should be crushed as coarse as permissible, and where the sulphides predominate largely, amalgamation in the battery is best avoided. The pulp discharged through a screen carries but a small percentage of the size of the orifice; while the largest proportion is much finer, it is possible to use a much coarser screen than the size desired to be obtained without any great detriment, while greatly increasing the output.

The Screen Frame (Figs. 24 and 25).—It is made from strips of sugar pine 1½" by 3" broad, mortised, and reversible; usually they are made to close the entire discharge opening, grooves being cast on the exterior of the mortar for their reception. It is frequently strengthened with one or more vertical ribs across the center opening, and is faced with iron plate on those portions of the side and bottom that come in contact with the iron keys that hold the frame solid against the mortar. In some mills the frame is made several inches lower than the opening, to permit the millmen to observe the interior of the mortar while in action, and to allow the hand to be introduced to remove any chips that may have passed in with the ore, as these have a tendency to bank up against the screen and interfere with the discharge of the pulp. Where such a screen frame is used the opening above is kept covered with a strip of canvas tacked to a wooden rod, laid on the upper projecting lid, while the loose end of the canvas hangs against the inside of the upper part of the screen frame.

The Plate-Block (*Chock-Block*) consists of wooden blocks bolted solidly together, and fitted and keyed to the lower edge of the mortar along the discharge opening, with one part projecting above the other, forming a recess on top to contain the screen frame, and lined with a piece of blanket to make a close joint. The inner side is sloped or rounded off, and fitted with an amalgamated plate. The front and ends are faced with iron plate to protect the wood. Two or more sets of these chock-blocks should be provided, of which one stands 2″ higher than the other; they are then used alternately, the higher one with new shoes and dies, to be replaced by the lower one when the dies are worn down somewhat, to retain a more even discharge than would otherwise be possible.

The Drop is the height through which the stamp is raised by the cam, and through which it drops when released. Usually it is the same for all the stamps in a battery, although the end and feed stamps sometimes receive a different drop. It is regulated through the raising or lowering of the tappet, and depends mostly on the hardness of the rock. It is one of the factors in determining the speed with which the blows from the stamp shall be repeated. The usual combination of the two in the California mills, is a low drop with rapid motion.

The Discharge is the distance between the top of the die when in place in the mortar, and the lower edge of the screen through which the pulp discharges. It is one of the most important factors in the duty of the stamps and the gold output from the ore. It should be maintained as nearly as possible at an even height through the entire period of crushing; the height of the chock-block or screen frame being lowered to correspond with the wear of the die. A further means used to retain an even discharge is by placing a 2″ iron plate under the dies, when worn thin. The discharge stands in a certain relationship with the fineness of the screen: low discharge goes with coarser crushing, a high discharge with the opposite. The discharge varies in California from 4″ to 10″.

Water Supply.—Water-pipes of 3″ diameter are brought along the front of the mortar near the upper edge, with branch pipes 1″ in diameter, supplied with faucets leading to the feed side of the mortar, to convey the battery water in at the back, or through the plank-covering on the top; this water is under moderate pressure. A second discharge pipe is carried down in front to the lower lip of the mortar, where a movable, perforated branch is turned across the front of the screen, discharging along the entire line on the lip; this second discharge pipe also supplies a hose. The battery water should enter both sides of the mortar in an even quantity, and the total amount must be sufficient to keep a fairly thick pulp that discharges freely through the screen. About 120 cu. ft. of water per ton of crushed ore may be considered an average, or 8 to 10 cu. ft. per stamp per hour.

Aprons and Apron-Plates.—The apron is a low table placed in front of the mortar, just below and in immediate proximity to the lower lip of the discharge, for the reception of amalgamated copper plates. It is set on a sufficient grade to permit the discharging pulp to flow over it in an even stream, while affording the suspended amalgam an opportunity to reach, and adhere to, the plate surface. The size, shape, and slope are at the will of the millman; but usually they are rectangular, with the plates screwed down to the table with copper screws, perfectly level and smooth, the sides being secured with wooden cleats. The grade given varies from ½″ to 2½″ to the foot, and the width of the apron is usually the width of the discharge-opening of the mortar. In some mills several of these apron-plates are placed consecutively, discharging from one to the other. They are usually rigid, but in some instances the apron next to the mortar stands on rollers, permitting it to be rolled back, and thus giving freer access to the front of the mortar. They should not be attached to the battery frame.

Sluices and Sluice-Plates.—These vary from 12″ to 20″ in width, and are placed below the aprons; they are usually set to a grade different from that of the apron. The plates can be fastened by cleats, or are laid overlapping at the ends, and, if not wider than 16″, do not need to be fastened down with side cleats; this permits of their being picked up and cleaned at any time without stopping the battery. The sluices are rarely over 16′ long—more frequently in lengths of 8′—and should always be placed double. The width and grade, as compared with the apron areas, are mostly faulty in California mills.

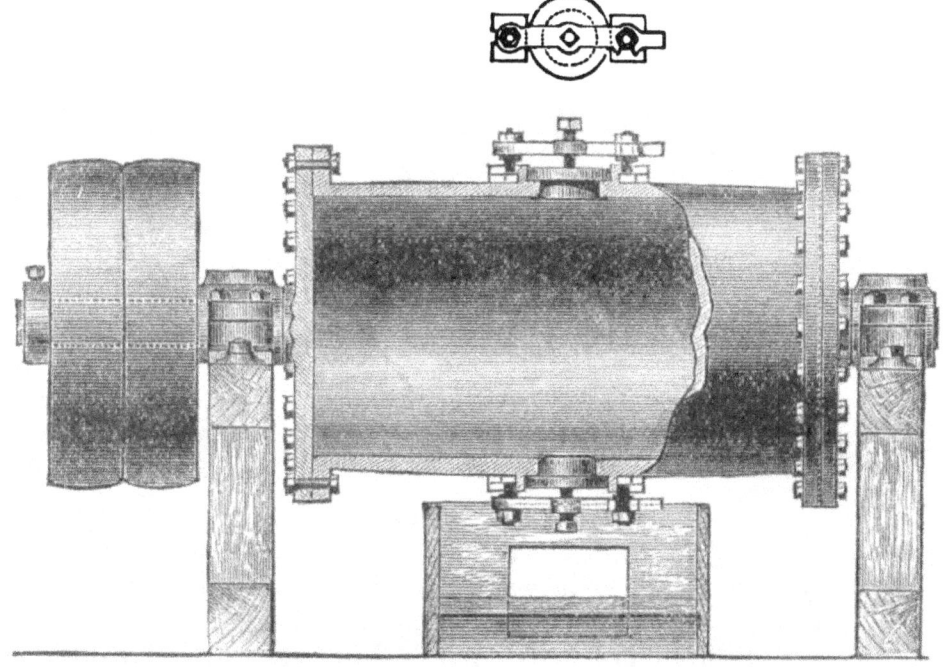

FIG. 26. CLEAN-UP BARREL.

Clean-up Barrel.—Large mills are supplied with clean-up barrels, which consist of iron barrels supported by trunnions resting in bearings on short standards. One of the trunnions is extended to carry a loose and a tight pulley, by means of which it is revolved. A manhole, with tight-fitting cover, is provided for charging and discharging, and below it is a sluice with cross-riffles to receive the pulp when discharged from the barrel. The barrel should make from thirty to forty revolutions a minute, requiring $2\frac{1}{2}$ H. P. It is used to treat the battery sands when cleaning up the mill; also, all the scrapings from the mill floors, as well as sand from the drop boxes and amalgam traps, large pieces of quartz or pieces of broken shoes being added with water and quicksilver to assist in the operation.

Clean-up Pan (Fig. 27).—This is a small amalgamating pan, 3′ to 4′ in diameter, operating with mullers with wooden shoes, and is run at a speed of thirty revolutions, requiring $1\frac{1}{2}$ H. P. When in use the pan is half filled with water, and the amalgam put in, with an addition of clean quicksilver, and, if required, also some lye. After sufficient grinding, the muddy water is run out through plug-holes, the mullers stopped, and the contents drawn off in buckets. The iron found floating on top of the quicksilver is removed with a magnet; the sand is washed off with a small stream of clear water, and if any dross be found covering the surface it is skimmed off with a sponge or piece of blanket.

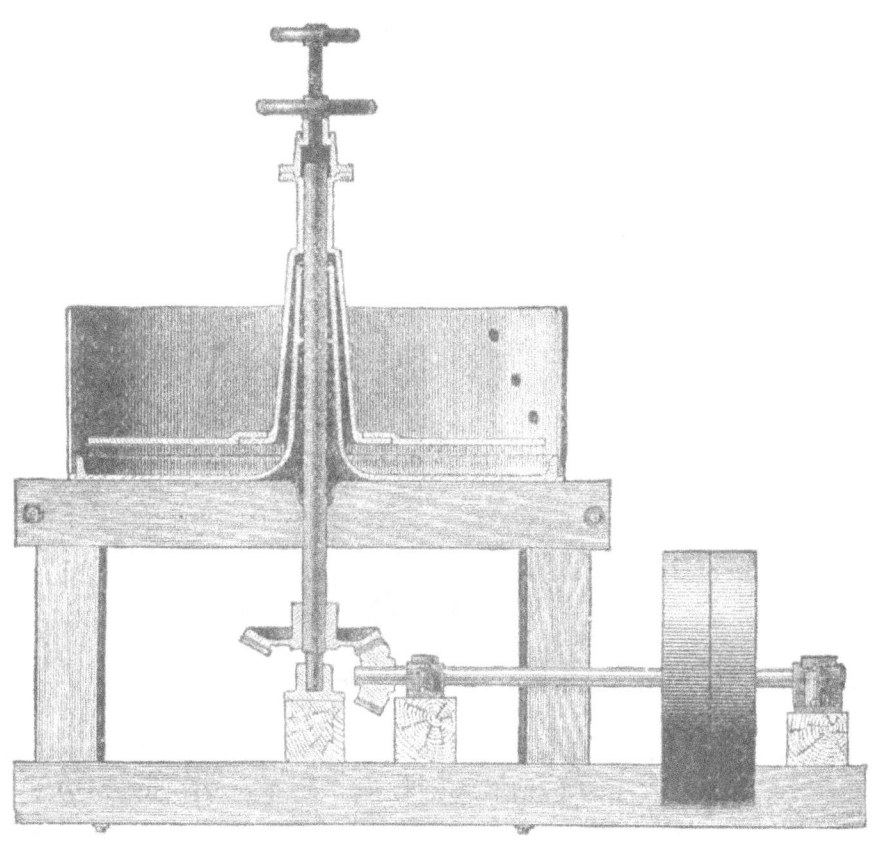

FIG. 27. CLEAN-UP PAN.

Cleaning-up Room.—This is an apartment in close proximity to the batteries and aprons, provided with a tight floor, and with a door under lock and key; the floor is best when laid in cement, to avoid all losses from split quicksilver or amalgam. It should be well lighted, and furnished with a sloping table large enough to place a screen frame on; also, with one or two water-tight boxes about 4′ long, 3′ wide, and 3′ deep, for panning-out purposes; these are supplied with plug-holes near the bottom, to drain off the water, besides water-pipes and fittings to fill the boxes when required. One or two wide shelves should be provided to hold the chemicals, quicksilver, and utensils needed in cleaning up. The latter consist of pans, wedgewood mortar, brushes, scoops, cups, knives, chisels, rubbers, scrapers, and a supply of closely woven drilling or light canvas; the latter is used to squeeze the superfluous quicksilver from the amalgam. A good pair of balances, with a set of accurate weights, capable of weighing the amalgam and the retorted bullion, should also be provided. The table should be made of a solid plank, or a slab of slate or marble, supplied with a raised edge, and grooved around to drain into a pan placed on a shelf attached below the lower end; some tables are covered with an amalgamated plate. It is sometimes convenient to have a small safe in the clean-up room, but it is always better to have the amalgam delivered to the office.

FIG. 28. THE KNIGHT WHEEL.

POWER FOR MILLS.

On account of the favorable position of the majority of California mines as regards their proximity to mountain streams and the large ditch systems, the application of water for the motive power of the mills is rendered easy, and where the distance from these sources is remote, electricity generated in such localities and transmitted to the mill is being successfully applied. Where steam power has to be used, the well-timbered western slopes of the Sierra Nevada permit the cost of fuel to be kept at a comparatively low figure. Where both water and timber are hard to obtain, as in the desert regions of the southern part of the State, gas engines have been applied with most satisfactory results.

In applying water power, where the pressure is sufficient, hurdy-gurdy wheels are chiefly used; these are vertical wheels with narrow breasts, having buckets of various patterns radially attached to the outer circumference, the water being projected through one or more nozzles against the buckets at a low point of the wheel, allowing the water to pass from the buckets as soon as the blow has been delivered. The principal patterns in actual use are the Knight, Pelton, and Dodds; the actual effective power developed by the Pelton buckets is given at about 75% to 80%. Where sufficient pressure cannot be obtained, the Leffel turbine and the overshot wheel are in use. As the Pelton wheel seems to find the most frequent application in California, it may be convenient for millmen to have the following rule, applicable to these wheels: When the head of water is known in feet, multiply it by 0.0024147, and the product is the horse-power obtainable from one miner's inch of water.

FIG. 29. THE PELTON WATER WHEEL.

The power necessary for different mill parts is:

For each 850 lb. stamp, dropping 6″ 95 times per minute _____ 1.33 H.P.
For each 750 lb. stamp, dropping 6″ 95 times per minute _____ 1.18 H.P.
For each 650 lb. stamp, dropping 6″ 95 times per minute _____ 1.00 H.P.
For an 8″x10″ Blake pattern rock-breaker _____ 9.00 H.P.
For a Frue or Triumph vanner, with 220 revolutions per minute_____ 0.50 H.P.
For a 4′ clean-up pan, making 30 revolutions_____ 1.50 H.P.
For an amalgamating barrel, making 30 revolutions _____ 2.50 H.P.
For a mechanical batea, making 30 revolutions_____ 1.00 H.P.

MILL PRACTICES.

Where the conditions permit, it is becoming the custom to place the grizzly and the rock-breaker in close proximity to the hoist, so that the bucket or car on arriving at the surface is dumped direct on a grizzly, and the crushed ore is then run over the ore-bin in the mill, and emptied therein; where this is impracticable, the grizzly and rock-breaker are placed over the ore-bin in the mill.

The usual practice is to let the coarse ore from the grizzly drop on a platform on a level with the mouth of the rock-crusher, into which it is shoveled by hand; by this method the machine is not brought up to its full capacity. A better plan is to convey the coarse ore from the grizzly into the bin by means of a chute, having a sliding gate immediately above the receiving point of the crusher, and which is set so as to keep the space between the jaws always filled. In this way the work becomes automatic, and the services of the man attending the rock-breakers can be utilized in other parts of the mill during part of the time. Under such an arrangement the crusher will require more power, which should be independent from the other machinery. The rock-breaker is usually run during the daytime only, as it can crush in that time enough ore for the mill for the twenty-four hours.

The self-feeders, in a similar manner, are kept automatically filled from the main ore-bin. The feeding through the tappet striking on the bumper-rod of the self-feeder, has of late been modified. A collar is fastened below the guides on the feed-stamp stem, taking the place of the tappet, thus avoiding the long bumper-rod. The gauging of the feed must be carefully attended to, if the stamps are to work up to their full capacity; there should never be more than about 1″ of rock between the stamp and die when they come together, or the feed should be just sufficient to keep iron from striking iron. When cleaning up the batteries, the self-feeders are drawn back on a track toward the ore-bin, giving access to the back of the mortars.

In preparing the mortar for ore-crushing, an inch or two of tailings is spread evenly over the bottom before putting the dies in place, as this saves the wear on the bottom plate. After the dies are placed exactly under each stamp, crushed ore and fine rock are banked around them to retain them in proper place until the sands have settled firmly about them. Care must be observed to keep the tops of *all* the dies at the same level at all times, as otherwise, when the stamps are dropping the highest die will strike against iron, while the others are still supplied with sufficient ore; this is known as "pounding."

The stamp-head, or boss, is now placed on the die with the small conical opening at the top, and the stem lowered into it, iron against iron if it is a close fit, and driven in solid. In case the connection is not tight, canvas strips about 2″ wide are laid crosswise over the opening before the stem is lowered. The stem, with the stamp-head, is now raised until the latch-finger catches under the lower face of the tappet and holds them suspended, and the shoe placed on the die. If the stamp-head hangs too low to permit of this, the stem is raised, and a block placed on top of the finger for the tappet to rest on. Narrow wooden wedges, about 1″ wide, the length of the neck of the shoe, and of the requisite thickness to fit tightly into the conical opening at the bottom of the stamp-head, are arranged in place and tied with a string. The block and finger are then removed, the stamp-head dropped over the shank, and wedges driven down firmly. This is done best by revolving the cam-shaft slowly, and, while placing the cam-stick between, permitting the cam to act on the tappet, raising and dropping the stamp until the lower edge of the stamp-head is nearly in contact with the shoulder of the stamp. It is not advisable to permit them to come solidly together, as it tends to loosen the iron ring that reinforces the stamp-head. A quick and convenient method of placing the wooden wedges on the shoe, is to cut a piece of canvas to fit exactly around the neck, and attach the wedges to the canvas by driving a tack through each one into the cloth. By keeping a supply of these on hand, it becomes an easy matter to encircle the shank on the shoe and tie them fast, should the shoe become loose and drop off while the mill is running.

The Drop.—The next operation is fixing the distance through which the stamp is to drop before striking the die. In most mills this distance is uniform for all the stamps; but, as previously stated, occasionally the stamp operating the feed, as also the two outside stamps, receive a greater drop.

The right height to give depends on the nature of the ore, as also on the speed to be given to the stamps; that is, the number of drops per minute. The tendency in most California mills is to run at a high rate of speed, usually in the neighborhood of one hundred drops per minute. The height varies from 4″ to about 10″, generally but little, if any, above the water-level in the mortar.

In arranging the stamps for an equal drop, wooden blocks, cut about ¼″ longer than the drop the stamps are to receive to permit the cams to clear the tappets, are placed on the die, between it and the shoe. Pieces of 2 x 4 scantling, cut to the desired length, answer well for the purpose. The keys in the tappet are loosened with a drift made of steel, the size of the keyholes, and used only for that purpose, and the stem is allowed to slip through the tappet until the shoe rests on the top of the wooden block beneath; or, if the shoe was resting on the block previously, the tappet is slipped up till resting on the latch-finger, when the keys are driven home solid. Care must be exercised not to drive the keys too solid, else there is danger of splitting the tappet. For the convenience of the millman, a chalk mark is made around the stem just above the tappet, which enables him, while running, to at once detect if any of the tappets have slipped. Should this occur it must be immediately re-set, or the battery work will be irregular. The battery-plates and chock-blocks are next put in place and keyed.

The Discharge is next arranged. This is the distance between the top of the new dies and the lower edge of the screen, and to fix the right distance is of importance. The greater the height of the discharge, the greater will be the proportionate amount of pulp and slime, and they also will be retained longer in the mortar. The quantity of amalgam retained in the mortar is also proportionately greater. A low discharge calls for a coarser screen, and naturally results in a larger output of the battery, and with a larger proportion of outside plate amalgam. With a *constant* height of the screen, the natural wear of the die increases the height of the discharge. For ordinary iron shoes and dies, and average rock, the wear of the die is roughly estimated from ¾ to 1 lb. of iron per ton of ore crushed. To counteract the effect of this wear on the discharge height, different-sized chock-blocks or screen-frames are supplied; the highest being used with new dies, and later replaced by lower ones, thus holding the distance more even than the use of a single size would permit. In some mills, when the dies are worn down, an iron plate, made for the purpose, is laid beneath them to raise them up.

As a very high discharge, besides creating much slime, beats up a larger portion of the gold into float gold than would be the case with low discharge, the choice necessarily influences the gold recovery; this is more particularly the case, if the ore carries any appreciable amount of valuable sulphurets. The discharge varies in the different mills from 4″ to 10″, the average being from 6″ to 7″.

Screens.—In fastening the screen to the screen-frame, care must be observed to get it on smooth, without any wrinkling or buckling. Tin screens must have the tin burned off before fastening to the frame; it is also well to expose the Russian-iron screens to a quick fire of shavings, to burn off the oil with which they are more or less faced. The edges of the screens are tacked to the frames, and are faced with strips of blanket to make a close connection with the mortar. In fastening a wire-cloth screen, to get it on smooth, a good method is to tack it first along the lower edge, then draw it up tight and even over the upper edge, and nail it before cutting it off the roll. As previously stated, brass-wire screens should not be used in conjunction with cyanide of potassium, as the brass becomes coated and clogged with amalgam. The screen-frame with screen is dropped into the grooves cast on the outside of the mortar discharge, and fastened solid with iron wedges—two vertical (one for each groove) and a horizontal one in the center of the lower lip. The wedges should have a broad head, to facilitate knocking them out. After the screen has been fastened in place, a piece of canvas or a board should be hung in front to arrest the outward throw of the pulp from the drop of the stamp, and direct it in an even flow onto the plates beneath. In some mills this board is given a slope toward the screen, and has an amalgamated plate screwed on, which receives the splash. Bolted to the front of the modern mortars is a frame to carry the outside battery-plate and a distributing-box, a few inches above the apron-table on which it discharges.

When everything is ready to drop the stamps, the self-feeder is rolled to its place, the cam-shaft is set to revolving slowly, the water is turned into the battery, and the millman, standing on the platform above, grasps the hand-hold of the first finger or prop and introduces, with the other hand, the cam-stick between the tappet and the revolving cam; by this means the weight of the stamp is taken off the prop, which is pulled back and rested against the edge of the platform. This operation is repeated with each stamp until all are working. To carry out this operation when the shaft is revolving rapidly, without injuring the operator's hands, requires practice. The cam-stick mentioned above consists of a piece of wood about $2\frac{1}{2}'$ long, $1''$ thick at the point, running up to $2\frac{1}{2}''$ near the handle, and faced with strap-iron or a strip of belting. It may also be made entirely of strips of belting, $2''$ or $3''$ wide, nailed over each other and attached to a wooden handle. To hang up the stamps, the hand-hold is grasped, the knee pressed to the latch-finger, and the cam-stick introduced between cam and tappet as before, and the latch-finger pushed under the tappet.

Before dropping the stems the face of the cams should be lightly lubricated, for which purpose axle grease or specially prepared compounds are used; a very useful one is a mixture of graphite and molasses; in some mills, to avoid the use of grease, the face of the cam is rubbed with a bar of common soap.

Grease.—It being essential, for good amalgamation, that the presence of grease be avoided in the battery, care must be observed in lubricating the cams, the stems where passing through the guides, and the shaft bearings. In many mills, trays made from old oil-cans are fastened beneath the bearings, cloth aprons are tacked from the underside of the guides to the floor above; rings of rubber packing or old belting also encircle the stems at the lower edges of the guides. The millman should diligently wipe off the stems and any part of the battery frame, where the presence of grease is indicated, at least once during a shift. Grease in the mortar is indicated by a black, dirty appearance of the surface of the plates, as also by the adhesion of more than the usual proportion of the amalgam to the iron castings inside the mortar. The usual remedy is to shut off a part of the battery water, for a short time, while adding a lye solution; or to add fine wood ashes to the ore.

The Amount of Water Required for the proper working of the battery depends on the nature of the ore; clayey ores, or such as have a high percentage of sulphurets, requiring the most; but while in the former case a greater amount is needed inside the mortar, the latter condition permits a part being added outside the screen, on the lip of the mortar. A small sluice-box, with plug-holes, is placed across the front in this case, or the water is conveyed by means of a half-inch perforated iron pipe, attached to the vertical supply pipe by an elbow joint, permitting it to be turned either way as required. "The amount of water used per ton of ore stamped varies from 1,000 to 2,400 gallons, with a mean amount of about 1,800 gallons per ton of rock crushed." * Most of the mills in actual practice figure roughly on one miner's inch of water, more or less, per twenty-four hours for each battery of five stamps. To obtain the largest amount of crushing of clean quartz from a battery, only sufficient water should be used inside to keep up the regular, even swash of the pulp, and if that be not sufficient to keep the plates on the outside clear from accumulating pulp, more may be added outside the the screen. The pulp, in passing down over the apron-plate, should roll in successive waves, corresponding to the back and forth wave-motion inside the battery, rather than flow in an even sheet, as affording a better opportunity of contact for the particles of amalgam.

Where the temperature falls low in winter, arrangements should be made to deliver the water in a tepid condition, as better amalgamating results will be obtained, through keeping the quicksilver in a lively condition. Where steam power is used, this can be easily arranged; but when using water power, a separate heater is required.

Feeding.—Hand-feeding has become nearly obsolete in California. It is only practiced in small concerns, or where a temporary mill has been put up for prospecting purposes. The advantages of a machine-fed mill are numerous; the chief of these are (1) that the wear of the iron of the shoes and dies is less and more even-faced; (2) that from 15% to 20% more ore can be crushed in a given time; and (3) that the labor expenses are reduced. The machines should be carefully gauged and watched to insure a steady, low feeding of the stamps. In order to insure a good splash in the mortar attention must be given to the succession in which the stamps are made to drop. A good splash is one that shows a wave passing along the lower edge of the screen, moving backward and forward from end to end, or a similar wave-motion that has its initial point from the center stamp. The succession most frequently adopted in California is 3, 5, 1, 4, 2; 1, 5, 2, 4, 3; 1, 3, 5, 2, 4, and 1, 4, 2, 5, 3; the last spreads the pulp very evenly from end to end. The greatest amount of discharge is obtained, apparently, by dropping the center stamp first; while the most crushing is done, other conditions being equal, by dropping the end ones first. Any arrangement of the stamps will answer, however, that distributes the pulp evenly and discharges it well.

The Apron should be set immediately in front of the mortar, but independent of the battery-frame, to exempt it from the jar of the stamps; it should be arranged to permit of the grade being easily altered if necessary. The size, shape, and grade of the apron-plates

---

* See VIIIth Report of State Mineralogist, p. 710.

differ widely, depending largely on the millman's preferences and experience. The usual form of the apron is rectangular, of the width of the discharge, and any length desired, but usually from 4' to 12', forming a level (transversely), smooth surface, set on a grade varying from $\frac{1}{2}''$ to $2\frac{1}{2}''$ to the foot. Sometimes the surface is divided by steps, with or without distributing-boxes. These are usually from 1'' to 2''. The apron should never be drawn in at the lower end, for reasons given farther on; and the steps should not be too deep, as otherwise the plate next to the drop will show mostly bare copper through scouring.

On examining a plate that is in use under good working conditions, it will appear that the upper portion, immediately below the mortar, say for a distance of 18'', carries at least 75% of all the amalgam caught on the apron, the largest accumulation showing along the line of impingement next the lip of the mortar. Now if the apron-plate were discontinued at about 2', and continued again on a lower level of about 2'', a second line of accumulation would result, naturally on a smaller scale; hence, the advantage of the step form. Another advantage in this style of apron, is that by fastening these sections to the table by means of wooden buttons on the sides instead of cleats and screws, and having one extra plate on hand, the scraping and dressing of the same can be performed at any time, without stopping the crushing of the stamps, by removing the plate and substituting the extra one.

The grade of the apron-plate should be such as to keep the surface clear from any pulp accumulations, but not steep enough to obtain a scouring action. It will depend on the coarseness of the pulp, the nature of the gold, the amount of water available, and the percentage and nature of the sulphurets. Where a battery-plate is in use above the apron, it is usually given a grade of from $1\frac{1}{2}''$ to 2'' to the foot. Grades for the apron proper vary from $\frac{1}{2}''$ to $2\frac{1}{2}''$ to the foot, but the average is about $1\frac{1}{4}''$. The apron-plates are usually silver-plated copper plates, which have largely superseded the copper amalgamated plate of former days—chiefly on account of the readiness with which the former plates do their full duty from the first starting, which is not the case with the copper plate; also, on account of their freedom from discoloration by oxidation. If silvered plates be used when running a very low-grade ore, the plating soon wears off, requiring a replating about every six months. The usual amount of silver put on plates is one ounce to the square foot. The usual thickness of copper plates is $\frac{1}{16}''$ to $\frac{1}{8}''$. In preparing them for amalgamation they should first be carefully heated to a black heat, and plunged into cold water, which makes them soft and more ready to take up quicksilver. They are then scoured bright with fine tailings-sand, moistened with some cyanide of potassium, and applied with a block of wood; then dressed all over with a weak solution of nitric acid, or with cyanide of potassium and quicksilver, with sodium amalgam sprinkled over and brushed or rubbed into the surface. Before final use, it is well to give them a coating of fine gold amalgam; or, if not convenient, silver amalgam will answer. In using the cyanide of potassium solution, care must be taken not to use it too strong, especially if the quicksilver is not applied to the plate immediately, otherwise a coating is formed on the surface that will not take up the quicksilver. Where the ore is of a fair grade, after a long period of continuous use the plate will have absorbed an amount of gold that will not yield to scraping unless the plate is immersed in boiling water for a time before being scraped, or heated over a fire and hammered with a mallet on the reverse side, in which case care must be taken not to dent the plate.

As the saving of amalgam on the apron and sluice-plates is largely a matter of gravity, the conditions under which the pulp passes over the plates should conform to the laws pertaining to the falling of a body through a moving liquid medium; hence the proper shape of the apron, and the flow and consistency of the pulp, should be well considered. If, as was formerly the almost universal custom, the lower end of the apron be contracted (and in numerous cases this contraction was as great as four to one), the depth of the pulp spread over the surface of the plate increases as it passes down; the flow of the water across a given section becomes uneven, forming at the sides a swirl, along the edge of which, sand is precipitated, covering and rendering that portion of the plate useless, from its inability to come in contact with the particles of amalgam, while producing scouring currents at other parts. The proper method is to spread the flow over a wider surface as it passes from one plate to the other, and lessen the grade, which may require an addition of clear water.

This contraction of the plates is made to this day in most of the mills, when connecting with the sluice-plates. The liquid pulp, starting with a width equal to that of the mortar-discharge, is made to pass over sluice-plates from 1′ to 2½′ in width; hence, the comparatively small percentage of amalgam obtained from them. The only condition under which narrower plates are permissible, is where, previous to receiving the pulp, a certain amount of the solid matter has been diverted. Where all the pulp goes from the plates to concentrators, the latter become an important factor in regulating the amount of water turned into the battery. The feed-water required for concentrators of the vanner types is from one to two gallons per minute.

In dressing the apron-plates prior to starting the stamps, they are first washed down with the hose, to remove all particles of coarse sand which might otherwise scratch the plate during the subsequent dressing, then rubbed with a brush, using, if necessary, some fine tailings-sand to remove all spots or stains. During this part of the operation, the brush is moistened with different chemicals, according to the preference of the millman; some use weak cyanide of potassium; others use strong brine, with a small addition of sulphuric acid; also, sal ammoniac, or soda, or lye, besides other combinations. In many cases these prescriptions are carefully guarded by their possessors as trade secrets, and are considered the basis of all the success the owner has achieved in his business. Anything that will give the plate a clean surface, free from oxidation stains, and retain for the quicksilver its bright condition, is useful in this respect. The main point to achieve success is to always keep the amalgam on the plate bright, and of the right consistency, and this art can only be perfectly acquired by actual practice around the battery and plates. After the plate has been thoroughly cleaned, quicksilver is thinly sprinkled over the entire surface, through a cloth, and spread evenly by means of a brush or piece of blanket, and finally the surface gone over with a soft broom or brush, from side to side; this leaves the amalgam remaining on the plate with fine ridges parallel to the screen.

Among the plate devices used in California mills, which may take the place of the apron-plates, or may follow them, is a late invention known as the *Gold King Amalgamator*. It consists of an iron cylinder, or drum, 6' long and 12½" in diameter, divided lengthwise into two equal parts, hinged together, and capable of being locked. Fitting tight inside of the cylinder are two corresponding semi-cylindrical silver plates, each with two longitudinal ribs set radially, at one-third distance apart and about 3" deep. The upper end of the cylinder is furnished, around the

FIG. 30. GOLD KING AMALGAMATOR.

circumference, with tooth-gearing, into which fits a spur wheel with a four to one transmission, driven by a 12" pulley. In the center of this end is a 3" feed opening, through which the pulp is dropped into the revolving cylinder. A trunnion at the lower end rests in a slide bearing, that permits of fixing the grade to be given the cylinder by means of set screws. The machine makes forty revolutions per minute, the pulp requiring about 3½ minutes to pass through the machine before being discharged. It is run by less than ¼ H.P., and is easily set up.

FIG. 31. GOLD KING AMALGAMATOR.

The pulp, when dropped in the closed cylinder, is caught by one of the ribs and raised to the highest point, when it drops, to be again taken up by the next rib, advancing at the same time a short distance ahead. The discharge is through the center at the lower end of the cylinder. From 15 to 20 tons can be passed through in a day; or for a larger sized machine, from 25 to 40 tons.

Where concentrators are used in the mill, the sluice-plates that follow the aprons are usually not over 8' in length and from 16" to 20" wide, with less grade than the apron.   This latter point is reversed in some mills, and the sluice-plates are comparatively steep.

Between the aprons and the sluice-boxes a drop box is placed, into which the pulp from the aprons discharges; there is one to each apron, or one for two adjoining ones.   These boxes are 1' wide and about 10" deep, with flat or partly sloping bottoms; these latter, generally where one box is used for two aprons, the bottoms sloping from each end across the width of the apron, toward a central part where the bottom is level, and from whence it passes by overflow to the sluice-plates.   These sluice-plates are in short lengths, and are either laid overlapping or screwed down to form a continuous sheet, and are prepared and treated in the same manner as the aprons.   Of late years a useful

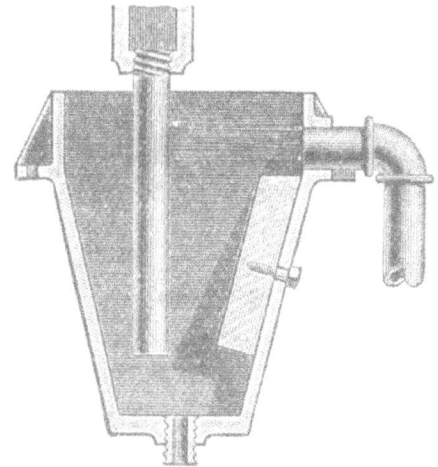

FIG. 32.   AMALGAM TRAP.

addition is being made to the plates in the form of a shaking-plate, of the same width as the aprons, and immediately below them.   It is either suspended or on a movable frame, and is given an end or side-shaking motion and light grade; for an end shake, the motion is imparted by a cam with $\frac{3}{8}$" stroke, and two hundred strokes per minute. The correct strokes for these plates must be determined at each mill by experiment.   Their efficiency was demonstrated in one mill, where the pulp passed over two consecutive apron-plates, and then to the shaking plate, which accumulated a greater amount of amalgam then the second apron.

Amalgam Traps (Fig. 32).—To retain any quicksilver or small particles of amalgam that escape inadvertently while dressing or cleaning the plates, traps are generally placed below the sluice-plates, and are made of various patterns.   The general idea is for the pulp to drop to the bottom of a deep vessel and flow out at or near the upper edge; in some cases, passing over a series of inclined shelves of copper plates during the descent.   A simple and very efficient contrivance for an amalgam trap is to suspend a narrow box by one end and attach the opposite end to a rod connected by a pin to an eccentric, through which it receives a gentle shaking motion in the direction of its long side.   The tailings are introduced into a stationary box immediately above, from whence, diluted with fresh water, the pulp passes over the top of a partition in an even sheet to the suspended box below.   The proper motion for this lower box must be found by experimenting, for which purpose the end of the rod is supplied with a series of holes, to shorten or lengthen the stroke. The motion should be just sufficient to keep the pulp suspended like quicksand, without splashing or caking on the bottom.

Amalgamating.—Quicksilver is charged by hand into the mortars through the throat, at stated intervals, with a small wooden spoon. Automatic quicksilver-feeders have been invented that are worked from the cam-shaft in such a manner that, at stated intervals, a little cup on a ratchet wheel, in revolving, dips quicksilver from a reservoir and drops it through a tube into the mortar. This insures absolute regularity; but for some reason they do not find much application in California. Retorted or new quicksilver should be used for charging as well as for dressing the plates. It is a good plan to keep the quicksilver used for these purposes covered with a weak solution of cyanide of potassium.

Quantity of Quicksilver.—To form some idea of the amount of mercury necessary to be introduced when handling an ore, the value of which is not known, a horn-spoon test of a weighed quantity is made, and the quantity of gold decided. Gold alloyed with an appreciable amount of silver requires a larger addition of quicksilver than does a purer gold. One ounce of gold of average fineness can be amalgamated with 1 oz. of quicksilver, but for a safety margin, an allowance must be made, so that 2 ozs. will answer better; and with extremely finely divided gold, 2½ or 3 ozs. If the stamps have a duty of two tons each, the amount of mercury requisite to amalgamate the gold contained in one ton of ore should be divided into five parts and introduced at half-hour intervals. If the ore be of low grade, the necessary portion may be added every hour; as the value increases, the stated intervals for charging should be reduced. The larger proportion of California gold ores receive mercury every half hour.

The skilled millman judges from the condition of his plates as to whether he is charging correctly. He places his finger on the apron-plate, and if the accumulated amalgam gives to a gentle resistance, and has a putty-like feeling, the condition is about right; when hard to move, he must increase the charge; or if thin, reduce it. The harder the amalgam, the more it assumes a dead-white color. The matter of correct charging of the mercury requires a constant watching, as on this depends the success of battery amalgamation; hence, the ore should be frequently tested with the horn-spoon.

Amalgam retained on the inside battery-plates weighs heavier, for the bulk, than the apron amalgam. There is a diversity of opinion among millmen as to how often the amalgam accumulated on the aprons and sluices should be removed. Thus it is found in the California milling practices that aprons are scraped as often as twice a day in some mills, while in others it is allowed to accumulate from one clean-up day to the next, which sometimes means once a month. Personal experiments by the writer, conducted in various mills, invariably showed a yield of more amalgam from the more frequent removal of the accumulations, but as the clean-up of the apron would require the cessation of crushing, such frequent stoppages would materially lessen the output. To avoid this, as the upper 18″ of the apron-plate retains about 75% of all the amalgam on it, this much of the apron-plate may be made separate from the rest, and held in place by wooden buttons on the side, so that it can be removed at any time while the battery is at work, and an extra plate, provided for the purpose, slipped in its place. Once or twice in the twenty-four hours it is advisable to hang up the stamps, one battery at a time, and dress-over the surface of the apron-plate, sprinkling, if necessary, a little fresh mercury, and brushing it into the adhering amalgam, after which the amalgam should be evenly spread out again. This takes but very few minutes. Frequently, when dressing a plate, a very fine coating of a brownish or grayish color can be seen adhering to the surface, which, on the application of the brush, is easily detached and thoughtlessly washed off. If this be examined under the glass, it will be found to contain considerable gold, hence should be gathered carefully in the gold-pan.

To remove the amalgam from the plates, the stamps are hung up, the battery-water shut off, and the front of the screen and plates hosed off to remove any sand which would scratch the plate. The surface of the plate is softened by the addition of quicksilver until the amalgam moves readily. Then, commencing at the bottom and working upward, with a piece of rubber, or rubber belting, 4″ long with square edges, the amalgam is pushed ahead to the upper end of the apron, gathered in a heap, and transferred to a pan or bowl by means of a scoop. The amalgam is taken to the clean-up room for further cleansing.

Where the amalgam has been retained on the plate for any length of time, as during an entire run, it requires a chisel or case-knife to remove it thoroughly, care being taken not to scratch the plate. In scraping a plate it is not advisable to remove ("skin") all the amalgam; enough should be left to form a thin coating, when ready to commence crushing again.

All mills experience more or less loss of quicksilver, partly through careless handling in dressing the plates, but also from the "flouring" of the mercury (breaking up into minute globules) after charging in the battery. This loss is extremely variable in the different mills, depending on the nature of the ore, high discharge, and low temperature of the battery-water. Ores carrying much talc, black oxide of manganese, galena, or arsenical pyrites, cause a good deal of flouring of the mercury. A further cause of loss is through incomplete retorting, a certain amount of mercury being retained in the bullion, which is volatilized in the subsequent melting. One half ounce to the ton of ore may be taken as near the average loss for California mills, although in a few cases these figures are doubled.

As the amalgam retained in the battery is less liable to loss than that portion adhering to the outside plates, the aim of the millman is to retain the largest proportion inside the screens. The coarseness of the gold has a good deal to do in this direction, as well as the splash and height of discharge. In some mills, as high as 80% of the total yield of amalgam will be found in the battery; it is always greatest, with the same grade of gold, where the most copper-plate surface is found inside the battery. The average proportions of amalgam retained in this country may be stated as two thirds in the battery as against one third on the (outside) plates, depending, of course, on the character of the gold in each district.

As the proper condition of the mercury is a matter of importance to the millman, it is well to become familiar with its different phases. Pure mercury is bright, quick, and does not change its appearance on exposure to the air at ordinary temperature, but evaporates slightly. As the temperature decreases it becomes stiffer, and at low temperature assumes a more leaden appearance; in raising the temperature it becomes more liquid. At 60° Fahr. it emits vapor sufficient to discolor a bright piece of gold when suspended over it in a closed vessel. Pure mercury, if dropped into a porcelain dish or on a table, will form into spherical globules, whereas the impure metal breaks into pear-shaped drops, and if very impure, the particles drag a tail when moved. If containing lead, a skin of metal will remain on the fingernails when passing the hand through the surface. The introduction of grease or unctuous substances, like clay and talc, incline the metal to separate into extremely fine globules—flouring. Quicksilver is attacked by heated concentrated sulphuric acid, but is not affected by it when diluted. Muriatic acid likewise does not affect it. Nitric acid attacks it and forms nitrate of mercury, a white compound. Quicksilver that has been used in gold-milling dissolves and retains a certain amount of gold, which remains with it, even after retorting. If quicksilver of this description be left for months undisturbed in a cold place and then carefully poured or siphoned off, a network of fine, needle-shaped crystals of amalgam will be found in the bottom of the vessel, derived from this gold held in solution.

Sodium Amalgam.—As sodium amalgam is frequently added to the quicksilver by millmen, the following method of preparing it is given: Dissolve small, dry chips of clean sodium, freshly cut from a stick, in pure, dry mercury, gently heated in a flask or porcelain dish; add it piece by piece until the mass has attained the consistency of soft putty, which should always be kept dry and well bottled, as it deteriorates rapidly in the air. This preparation is added to the mercury when dressing the plates; and to know when the proper amount has been added, dip a brightened nail into the quicksilver, which will adhere slightly to the edges of the nail if the amount be correct; whereas, if it becomes entirely coated, too much has been used, and more quicksilver must be added; on the other hand, if there be no signs of adhesion, more sodium amalgam must be added.

Nearly all commercial mercury needs cleaning. The handiest way is to digest with dilute nitric acid for twenty-four hours, taking one part of acid to three of water. In retorting foul quicksilver to purify it, the retort should only be half filled and the quicksilver covered with a layer of quicklime or charcoal powder. The heating should then be done very gradually, the retort not being brought to a full red heat.

Cleaning Up.—When ready to clean up a mortar, the feed of ore is shut off, and the speed of the stamps reduced until as much of the sand, etc., as possible has been discharged and iron strikes on iron. The battery-water is then shut off, the self-feeder pushed back, the stamps hung up, the splash-board or canvas removed from in front of the screen, and the face of the latter washed off with the hose. The aprons and plates are then scraped, and the aprons, if fixed, covered with planks near the mortar, to protect them while working around the mortar. The keys that hold the screen in place are withdrawn and the screen-frame loosened and slightly raised, permitting the water that is still retained in the mortar to gradually run out; a too sudden raising of the screen-frame from the chock-block would cause the water to escape in a body and possibly wash amalgam from the plates. After raising the screen out of the grooves, the chock-block and inside plates are removed and all of them carefully washed over the apron, scraped, and set to one side or removed to the clean-up room for treatment. The sand mixed with ore on and around the dies is taken out by trowels and passed through some other mortar, or retained to place around the dies when returned to the mortar. The dies are broken out of their beds with the help of chisels and crowbars; when the center or end die has been successfully worked loose, the remaining ones are easily taken out, washed, examined for any adhering amalgam (which is scraped off), and placed on the floor, *in the same order* they occupied in the battery, ready to be replaced. The remainder of the material in the mortar is then easily removed, and placed in the clean-up barrel; in small mills it is panned in a water-box provided for the purpose in the clean-up room. In the revolving clean-up barrel, pieces of quartz or old iron, with an additional amount of quicksilver, are added, and the barrel is half filled with water, where it is left revolving for a couple of hours. As all battery sands contain more or less nails and chips of iron and steel, these are removed by a magnet while panning out. The clean-up barrel is discharged through a manhole into a bucket placed over a riffled sluice. The bulk of the quicksilver and amalgam is retained in the bucket, and the overflow passes into the sluice.

After all the sand, etc., has been removed from the battery, the inside is washed out, and any amalgam found adhering to the sides or linings is carefully scraped off with a case-knife and placed with the rest of the amalgam for further cleaning. A bed of dry tailings-sand is then spread over the bottom of the mortar, and the dies *replaced exactly* as they were before. The tappets are then set, plates and screens put in, the feeder replaced, water turned on, and the battery once more started.

The operation of cleaning-up the batteries is performed usually once or twice a month, and in some mills once a week, at which time tappets are re-set and any necessary repairs made; also, any shoes that are too thin or broken are knocked from the boss and new ones substituted. As one new shoe in a battery of old ones causes irregular working, it is best to replace all the shoes at the same time, and if any of them are not worn down thin enough to discard, they may be set aside and used to replace a broken one at some future time. The same thing holds good with the dies, for if they are of uneven height they interfere with the regularity of "splash," and the higher die will be pounding iron while the remainder have still a sufficient cushion of quartz.

The amalgam obtained from a clean-up is washed in small batches in the gold-pan, to free it from all sand, fine iron, or sulphurets, and then stirred up with an excess of mercury in a wedgewood mortar, bringing all impurities to the surface; this dross is skimmed off and collected for further cleaning. The superfluous quicksilver is squeezed through a straining cloth or closely woven drilling, or through buckskin, and the resulting balls of amalgam retorted. This squeezing is best done by hand. After first thoroughly wetting the cloth or skin, it is laid loosely over a cup or bowl, and a convenient amount of amalgam poured in the center, enough to make, when squeezed, a ball of 20

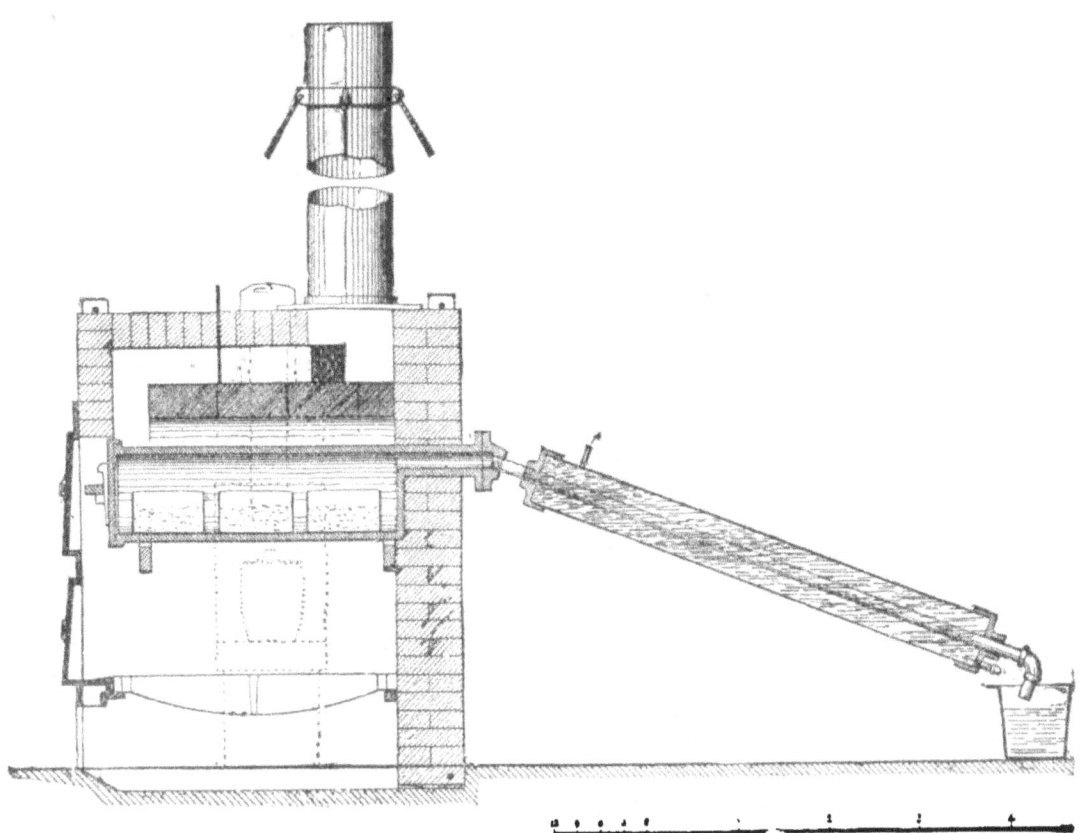

FIG. 33.

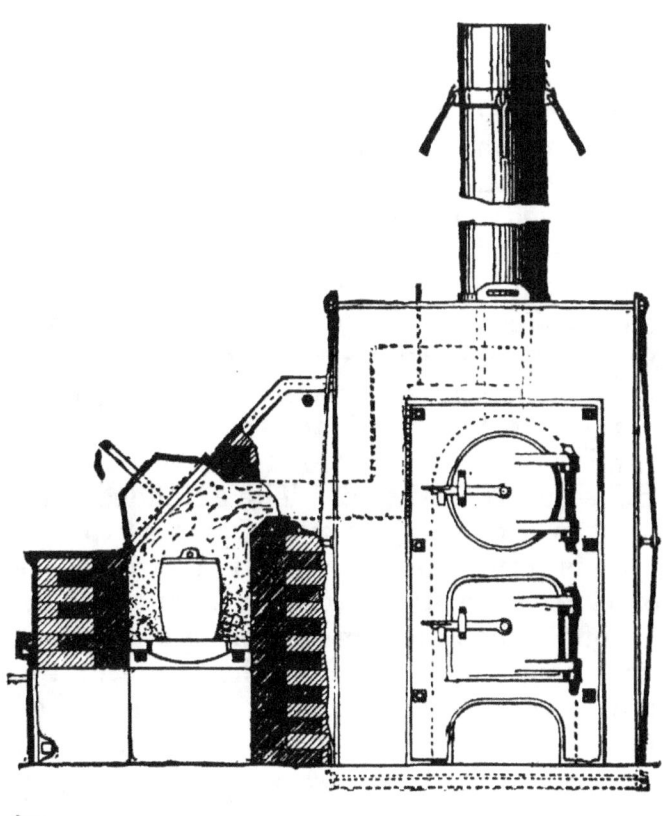

FIG. 34.

to 30 ozs.   The ends of the cloth are then gathered tightly together, and commencing near the ends, it is twisted until the amalgam is compressed to a hard ball, the strained quicksilver dropping into a pan of water beneath.   It is not good practice to squeeze the balls too dry, as the last quicksilver expressed is heavily saturated with gold.

In large mills the retorting is done in pans placed in an iron cylindrical retort built into a furnace, where the flame passes under and around it.   (See Figs. 33 and 34.)   But in the majority of cases in California they use the cup-shaped iron retort.   (See Fig. 35.)   These

FLAT TOP RETORT.

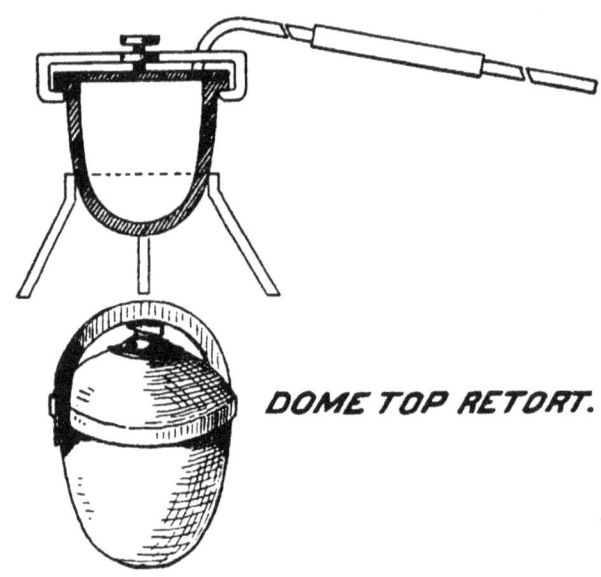

DOME TOP RETORT.

FIG. 35.

are made in different sizes, numbered from 1 to 7; No. 1 containing 150 ozs., and No. 7, 2,000 ozs.   They are made of cast-iron, with flat or half-spherical lids, which are secured to the retort by clamps and wedges or thumb-screws, the flanges being ground together.   From a vent-hole in the cover a curved condensing-pipe, securely screwed in, extends several feet.   The retort is placed in a ring-standard, or suspended when retorting, and should always have a space of about 6″ beneath it.   In preparing to retort, the inside is well rubbed with chalk and the balls of amalgam broken up and dropped in loosely; not pressed down into a solid cake, as is sometimes the practice, as that retards the operation. The flanges of the retort and lid are then luted together with a thin paste of flour and water or sifted wood ashes and water (the former is preferable), and securely fastened.   The extended end of the condensing-pipe is placed in a vessel with water, and as this pipe must be kept cool, fresh water is kept passing over it during the entire operation. The retort should never be filled to its full capacity, to avoid danger of an explosion through the amalgam swelling and closing the vent.   At first a light fire should be started at the top, and the heat gradually increased until drops of quicksilver issue from the end of the condensing-pipe.   The retort should then be kept at a red heat until no more quicksilver is seen to issue from the pipe, when the temperature should be raised to a bright "cherry heat" for a short time.   The retort should be kept covered by the fire during the whole operation.   If during the retorting the condensing-pipe should suck water, it should be raised momentarily out of the water to permit of the latter flowing out. A better arrangement, and one that obviates this difficulty, is to attach firmly to the end of the pipe, a rubber or canvas bag in the water, which will distend itself as soon as the mercury commences to flow, and collapse when the distillation ceases. When the operation is completed, which usually occupies about two hours, if the amount be not very large, the quicksilver is removed and the retort taken from the fire and allowed to cool; the lid is removed and the retort turned over a *dry* gold-pan. If the gold adheres to the retort, a few taps with the hammer on the bottom or the help of a long-handled chisel will release it. Well-cleaned and retorted amalgam should show a good yellow color. If black spots be seen it is proof that the cleaning was not thoroughly done, and a pale-whitish color shows that it still contains quicksilver. Care should be observed, when removing the lid of the retort, to avoid inhaling any fumes retained therein. All retorted amalgam should be melted and run into a bar, before shipping, as it saves losses incurred by abrasion where the distance is great to the shipping point. The melting is performed in a black-lead crucible, which, when new, must first be dried and annealed by placing the inverted crucible and lid in the furnace with a slow fire, which is gradually increased until the crucible is red hot. When ready to commence melting, the crucible is placed on a firebrick in the furnace, after introducing the retorted bullion, in not too large pieces, with borax, and covered with the lid, adding, if necessary, more of the bullion as the metal subsides. After all is melted down, the slag is skimmed off carefully from the top of the metal, which should show a bright surface. It is then ready for pouring. Should the surface not appear bright, but show a scum on top, some lumps of borax must be added, the crucible again covered and heated, when the scum will be slagged and skimmed as before, when it is ready to be poured into a mould. Should the second addition of borax fail to produce a bright surface, a very little niter may be added with the borax. Before using the mould it should be warmed and smoked on the inside by holding over the flame of a lamp or over a dish with burning rosin. The metal in the pot should be stirred before pouring; the stirrer, an iron rod, must be heated before introducing it. The bar, when solid, is turned out of the mould, and any adhering slag is hammered off; it can then be dipped into water to thoroughly cool it, dried, and weighed. Two small chips should then be taken—one from an upper corner, the other from the diagonally opposite lower corner—to be assayed.

ASSAYING AND SAMPLING.

Although at present most California mills have their own assayers to test the ore and the tailings, the time was not so very remote when it was not considered requisite to do *any* assaying. The expert millman could tell (?) by horn-spoon test how much his ore would mill to the ton; and if a horn-spoon test of the tailings showed no amalgam, he confidently asserted that all was being saved. It was decidedly a case where "ignorance was bliss." No gold milling can be carried on understandingly without light being thrown on the different results achieved, and which can only be given by careful sampling and assaying. It is not sufficient to know that a certain loss has been sustained. It should be accompanied by a knowledge in what particular part of the operation the loss has been incurred, to enable the operator to remedy the evil; hence, the necessity of constant sampling and assaying. In some cases the loss will be found entirely in the coarse sands, indicating that the screens are not fine enough; again, the loss may be entirely through sliming of the ore, or the missing percentages of gold will be found mostly in the sulphurets. The assay test alone, *with correct sampling*, furnishes the knowledge.

Sampling.—Samples should be taken regularly of the ore as it comes to the mill, as well as of the tailings *as they pass off*, for without the knowledge derived from these two operations there is no means of controlling the work.

Ore, as it arrives at the mill, is sampled by taking a stated amount (shovelful) from each ore-car or wagon, and throwing the samples together in a pile on a clean-swept floor or into a small bin. The pile should be shoveled over after breaking the pieces to the size of macadam; or if the pile be too large, cut through it at right angles, throwing the rock from the trench thus made in a pile by itself. This should be crushed or broken to a nearly uniform size, mixed by shoveling, and made into a low, truncated cone, which is divided into four equal parts by making a cross on the surface, and throwing out two diagonal quarters, which are again reduced in size, made into a second similar cone, and treated as before. This quartering and crushing is continued until a half-pound sample is obtained for fire assay. Great care must be observed when removing the different quarters to see that all the fine dust is swept up and added to the pile each time, as otherwise very defective results will be obtained. The rest of the ore is returned to the main ore-bin. Samples taken in this way from the aprons of the self-feeders are likely to give a more correct average, having been crushed, and the coarse and fine duly mixed. Samples should be taken at regular intervals from the pulp *with the water* that has passed over all the plates, and also from the concentrators.

Tailings samples should be taken at stated intervals by passing a vessel across the entire width of the discharge, where they leave the mill, without permitting any to flow over, and gathering at each interval an even amount. This is allowed to settle in a bucket and the clear water then poured off carefully or drawn off with a siphon. The residue is dried and thoroughly mixed, and several packages of 5,000 to 10,000 grains each weighed out. In some mills tailings samples are obtained automatically, using their current as the motive power for the sampler, which works by intermittently deflecting a spout through the tailings where they finally drop from the sluices, obtaining the sample across the entire section of the current. An appliance for this purpose is shown in the accompanying drawing. (Fig. 36.)

To ascertain the amount of slimes in the tailings sample, put one of the packages into a bucket, add water, and stir it. After settling two or three minutes, pour off the muddy water into a separate vessel; repeat this operation until the water comes off clear; add a little powdered alum in the vessel containing the muddy water, and when the mud has all settled, draw off the top water carefully and evaporate the remainder. Dry the washed sands of the sample, and pass through different sized screens, weighing the different amounts as they have passed, and assay each size; this will show where the greatest loss is sustained.

To ascertain where the loss in sulphurets occurs, it is sufficient to pass one of the 10,000 grains samples through a 60-mesh wire screen; weigh that which passes through and that which remains on the screen, and

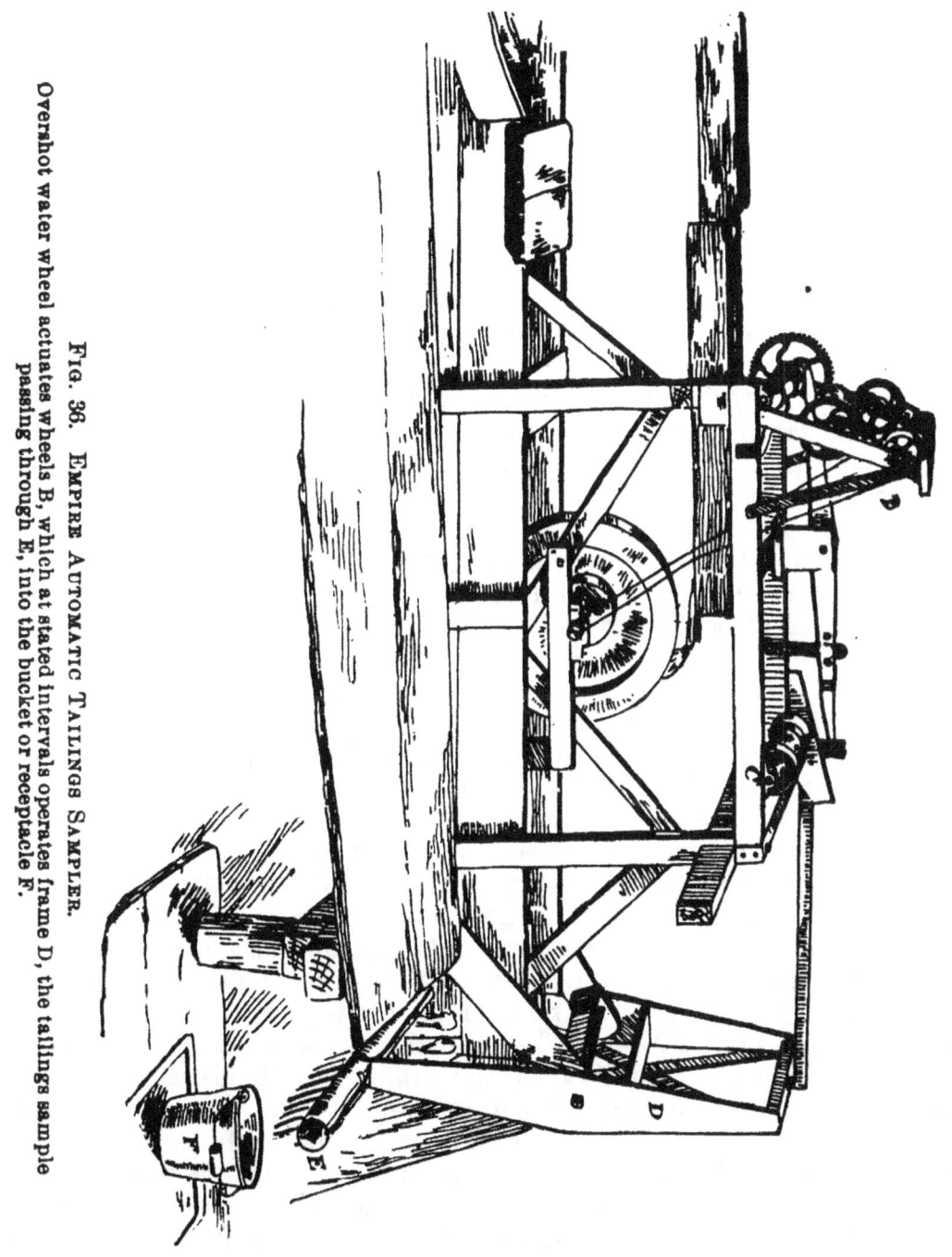

Fig. 36. Empire Automatic Tailings Sampler.

Overshot water wheel actuates wheels B, which at stated intervals operates frame D, the tailings sample passing through E, into the bucket or receptacle F.

pan out each lot carefully by itself, from one pan into another, as long as sulphurets can be recovered; then weigh each batch of sulphurets separately.

The use of 10,000 grains is recommended, as every 100 grains is 1%, and each grain is $\frac{1}{100}$ of 1%; it is also a convenient size for obtaining accurate results. By using pulp samples instead of tailings, the amount of sulphurets in the ore may be ascertained.

If the sulphurets assay $75 per ton, and the quantity per ton is 1.7%,

4—GMP

the value of the sulphurets in one ton' of ore is found by multiplying $75 by 0.017, which would be $1 27 per ton. If the loss of sulphurets in the tailings is 11 grains out of the 10,000 grains sample, and the value of the sulphurets is $75 per ton, then multiply $75 by 0.0011, and the value of sulphurets in the tailings is found to be $0.0825 (8¼ cents) per ton of tailings.

MILL ASSAYS.

Amalgamation (Free-Gold) Assay.—Take two pounds (being exactly one thousandth part of a ton) of ore, crush in an iron mortar, and pass through a No. 60 sieve; remove the gold and other metallic substances left on the sieve, and place in a small porcelain dish containing a little dilute nitric acid, to remove any adhering crusts of oxide of iron, etc., which might prevent amalgamation; these residues are then carefully washed and thrown into the sifted ore, which is then placed in a wedge-wood-ware mortar and mixed with enough warm water to make a stiff paste. To an ounce Troy (480 grains) of new, clean mercury, *free from gold*, add a piece of clean sodium about the size of a pea. The mercury thus highly charged with sodium is then thrown into the mortar containing the sample, and the mass ground constantly for an hour, when amalgamation should be quite complete. The mass is then transferred to a gold-pan and *carefully* washed over another pan or tub, in which the tailings are caught, and re-washed to save anything which may have escaped. The mercury is collected and transferred to a small dish; if it be much floured and refuse to run into globules, stir it with a small piece of sodium held in the end of a glass tube, which will cause it to run together. The mercury is then washed carefully in clear water and dried with blotting paper. It is then re-weighed, and if the loss exceeds 5% the assay must be rejected and a new one made. The mercury is next transferred to a small annealing-cup or crucible, which has been carefully black-leaded inside, covered with a porcelain or clay cover, and volatilized with a gentle heat. When all the mercury has been volatilized, about 50 grains of assay lead are thrown into the crucible and melted, giving it a rotary motion while in a molten state. It is then removed, cupelled, and the "button" weighed. It may be assumed without sensible error that the mercury lost in the operation carried the same proportion of gold as is contained in the mercury recovered; hence the gold contents of the ore will be found by multiplying the weight of the "button" obtained by the weight of the original quantity of mercury, and dividing the product by the difference between the weight of the mercury recovered and the "button." This figure, multiplied by 1,000, gives the weight, in grains, of the free gold and silver per ton of ore, which, for all practical purposes, may be assumed to be all gold. Should, however, greater accuracy be desired, hammer the "button" flat and thin, and dissolve the silver from it with nitric acid, and weigh the gold. The difference in weight represents the silver.

Panning Assay.—Take 2 lbs. of ore, crush, and pass through a No. 40 sieve; any gold in the residue left on the sieve being set aside. The sample is then carefully panned, and the tailings re-panned, to make sure nothing is lost. This operation will show at once whether the ore is rich in sulphurets or not, and the nature of them. The visible gold should be panned as free as possible from all the sulphurets, taking care that none is lost. The pan and its contents, together with the residue left on the sieve, are dried by holding over a fire; the contents are brushed into a cone of lead-foil, rolled up, melted, and cupelled. The "button" is weighed, and the free gold determined by multiplying its weight by 1,000.

The tailings produced in the panning operations should be panned over several times to collect all the sulphurets, which should then be dried, weighed, and their percentage in the ore determined.

Another method consists in not separating the free gold from the sulphurets, but in treating them both together by fire-assay, and determin-

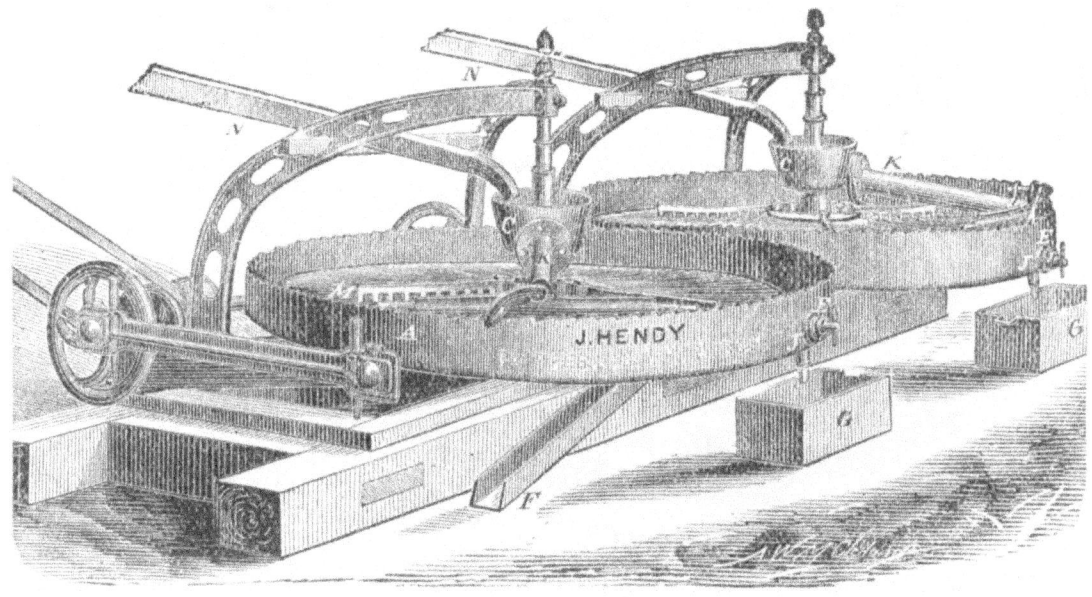

FIG. 36a. THE HENDY CONCENTRATOR.

ing the *total* value of the gold present in them. The operations, as far as described, are all that can be properly considered as coming under the term of battery amalgamation as practiced in California, if we except the use of the riffle and blanket sluices; these are placed below all the plates, and receive a very spasmodic attention in the majority of mills. Blankets are laid in strips, about 16″ wide and about 6′ long, overlapping each other in double sets of sluices, set on a grade of about ¾″ to the foot, washed in a separate water-box. The material thus obtained, with the contents of the riffles, is deprived of its valuable contents by the aid of arrastras, pans, or Chili mills. But few blanket-sluices are found to-day in California mills.

On the practical development of the Plattner chlorination process, by Mr. Deetken, in the "sixties," it was demonstrated that many of the low-grade quartz veins carried enough gold in their sulphurets to make their working profitable, causing attention to be directed to the concentration of these ores by mechanical contrivances. From the constant and successful use of the gold-pan the mechanical application of a similar motion was sought, resulting in the use of the Hendy and similar concentrating machines.

The Hendy concentrator consists briefly of a shallow iron pan with an annular groove on the outer edge and a waste discharge in the center. It is supported on a central upright shaft passing through the center of the pan, on which revolves, above the pan, a central bowl to receive the pulp, having two tubular arms extending close to the outer edge of the pan; these uniformly discharge the pulp at right angles from their axis. At a point on its circumference the pan is attached to a crank-shaft, making about 220 revolutions per minute. The sulphurets and small balls of amalgam gather in the groove at the outer edge, from whence they are drawn through a gate, which is regulated to be automatic in its discharge. This gate is not opened until the groove is pretty well filled with sulphurets. Two of these machines, driven by one shaft, are required for a five-stamp battery. The machine needs constant attention; one man can attend to twelve machines on a shift. They have been mostly displaced by the endless-belt machines which have developed from the endless blanket and shaking-table.

In 1867 the first patents for the revolving belt were issued.* This was the commencement of the belt concentrators, of which at present

FIG. 37. THE FRUE CONCENTRATOR.

the Frue, Triumph, Woodbury, Tulloch, Embrey, and Johnston are representatives. To produce the best results on these machines, all the stuff should be *sized*.

The Frue Vanner (Fig. 37), which has the largest representation in California gold mills, has been frequently described.† It has a side shake of 1", with from 180 to 200 strokes per minute, the belt traveling upward on an incline from 3' to 12' per minute. The belt is made in two sizes, 4' and 6' wide, and in the latest patterns as made at the Union Iron Works, San Francisco, has practical arrangement for easily chang-

* From the records of the United States Patent Office.

No. 61,426, January 22, 1867. T. D. & W. A. Hedger, Meadow Lake, California. "Revolving sluice for saving metals."
* * * "The endless apron is made of fabric sufficiently coarse to retain the heavier particles which it receives from the feed spout, beneath which issues a stream of water."
* * *
Claim 3. * * * "Separating the ore by passing the valuable portion up the incline and the debris down to the foot, as waste matter, as described." * * *

No. 66,499, July 9, 1867. George Johnston and Edwin G. Smith, Auburn, California. 'Amalgamator and concentrator."
* * * "The pulverized ore or tailings passes to an endless traveling and shaking canvas belt, which ascends against a stream, carrying the heavier particles to be discharged into a box, while the lighter ones are carried off." * * *
Claim 1. The revolving belt or apron (F), with its raised edges (O), having a shaking or rocking motion from side to side, substantially as used for the purpose herein described.

No. 239,091, March 22, 1881. Judson J. Embrey, Fredericksburg, Va. "Ore concentrator."

† See VIth Report of State Mineralogist, p. 92, article on Concentration, by J. N. Adams, E.M.; and VIIIth Report, p. 718, "Milling of Gold Ores," by J. H. Hammond, E.M.

ing the slope at the upper end. The frames of these modern styles are made of iron instead of wood. The pulp is discharged very evenly over the belt from a distributor near the upper end, just below the point where clear water is discharged in fine jets across the belt. In placing the machine, attention must be given to the solidity of the frame, and that a perfect level be obtained across the belt; further, the pulp and clear water must be distributed in an even depth of about ¼"; the grade and upper travel depend on the fineness of the pulp, and must be regulated accordingly.

The following guide for a proper condition of the work on the belt is given by Henry Louis, E.M., F.G.S., etc., in his very useful work, "A Handbook of Gold Milling," 1894, p. 324: "The working conditions should be so adjusted that a small triangular patch of sand should show at each of the lower corners of the belt. These sand-corners should not be too large, but must be well marked, and the two should be of equal size. Should they be unequal the fault will be found to be either in that the belt is not accurately level across, that the distributor

FIG. 38. THE IMPROVED TRIUMPH CONCENTRATOR.

is not doing its work properly, or that some of the working parts have not been properly tightened up, so that there are other motions than the normal ones communicated to the belts. Too large a corner of sand shows that the pulp is too thick, while absence of any corner indicates that it carries too much water."

Two of the 4' belt vanners, or one of the 6', handle the pulp from a five-stamp battery. The amount of clear water required to be added is about ⅕ cu. ft. per minute; the vanner requires about ½ H. P

The Triumph differs from the Frue, principally, in that it has an end shake of 1" and slightly quicker stroke (230 per minute), the belt making a forward movement of 3' to 4' per minute. It receives the pulp in a bowl containing quicksilver before reaching the distributor, which is all kept in agitation by revolving stirrers.

The Woodbury is similar to the Triumph in extent and number of motions, but divides the belt into seven longitudinal partitions; an increased output being claimed for this construction.

The Tulloch gives a rocking motion from a fulcrum on the floor, making 140 shakes of 1⅜" per minute, using either canvas or rubber belt. This machine, it is claimed, saves a somewhat larger amount of the finer and richer grade of sulphurets as compared with the former types.

The Embrey is similar to the Frue, but with end shake.

The Johnston, with improvements, and the latest of the belt concentrators placed on the market, claims many points of advantage. It is suspended from four non-parallel hangers capable of adjustment, by which the angle of oscillation can be changed as required, preventing the accumulation of sand at the edges, such as occurs with the horizontal side-shake machines, or the piling of the sands in the center of the belt, that occurs with the rocking motion. The motion imparted to this belt resembles more nearly that of the batea than that of any of the other concentrators. The belt is made of No. 6 duck, oiled and painted, but a rubber belt can be used at one third the cost of those with molded edges, which are short-lived. Small, hollow, brass, side-rollers on the shaking-frame, form the raised edges by curving the flat belt slightly upwards. The pulp is delivered from five slots running parallel with

Fig. 39. The Johnston Concentrator.

the belt frames, $\frac{1}{4}''$ wide and 16″ long, leaving 10″ spaces, into which the pulp is thrown when it strikes the belt. Here the separation at once takes place; the sulphurets settling on the belt are carried by it up to the clear water, while the sands are carried down the belt. In neither case are the sands or sulphurets obstructed by the falling of water and sands, as in other machines where the pulp is discharged across the belt. The clear water at the head of the table, instead of being discharged from a stationary box to the moving table, is discharged from a distributor, which is attached to and moves with the table, thus stripping the belt of the smallest possible portion of sulphurets. Two widths of belt, 54″ and 72″, are used, which are given a grade of $\frac{1}{5}''$ to $\frac{1}{4}''$ to the foot, making about 118 side-shakes per minute. One machine handles the pulp from a five-stamp battery.

Another vanner, soon to be placed before the mining public, consists of the essential features of the vanner, but carries a rubber belt with depressions all over it, 2″ in diameter and $\frac{1}{8}''$ deep, shaped after the batea, while the entire belt receives a motion corresponding to that given to a batea.

As the motion and grade given to any of these machines can only be correct for a certain size of grain in the pulp, it would be advisable to introduce some method of sizing the pulp previous to bringing it on the concentrator, and feeding the sized material to different machines. The finer the screen that has been used in the battery, however, the less does the lack of sizing affect the product from the concentrators. The concentrators should always, where possible, be attached to power independent from the stamps, and be placed on a floor below the aprons and in a position to permit the attendant to pass all around and to conveniently transport the concentrated stuff to the covered drying floor, which should be made with a slight incline, preferably of concrete, and exposed to the sunlight.

Canvas Platforms or Tables.—Investigation proving that the slimes passing off with the waste from the mill and concentrators still carried an appreciable amount of precious metal, millmen during the last few years have extended their operations, and re-treat the hitherto escaping slimes. This is done by conveying all the waste material from the mill, through sluices, to canvas platforms having the following general features. (See illustrations in chapter on "Typical Mills," pp. 66, 69, and 75.)

A platform is built of clear, seasoned, and planed, 1¼" planking, on a solid, level foundation, and given a grade of about ¾" to the foot, over which No. 6 canvas is stretched smooth, longitudinally, though sometimes crosswise, with a 2" overlap. Particular attention must be paid that the canvas is stretched smoothly and evenly and that no crack opens between the planks constituting the platform. The length and width of the platform required, depends on the amount of pulp to be handled; overcrowding must be avoided. The platform is divided longitudinally into sections corresponding to the width of the canvas, which is 22"; the partition is made of wooden strips, 2" wide and ½" high, covering 1" on the edge of two adjoining pieces of canvas. Running along the head of the platform are two sluices, one placed above the other; one containing clear water, the other pulp from the mill, both furnished with ¾" to 1" plug-holes over each section. Below the lower edge of the platform are two sluices placed side by side, the inside one to convey the waste, the outer one for the concentrates (sweepings) from the platform. When ready for operation, the plugs are withdrawn, and both pulp and clear water commingled flow down in an even current and are discharged through the bottom waste sluice. After one hour or less, the plug is inserted in the pulp-box over the first section, and the clear water permitted to run for a few minutes longer, during which time quartz sand may be observed passing off the canvas, leaving a dark, partly metallic-appearing sediment on the canvas. A tray or board is then placed over the waste sluice, connecting the lower edge of the section with the outside sluice, and the sediment is removed from the canvas, either by sweeping or with the aid of a hose with a flattened nozzle, to be worked later by chlorination or cyanide process.

The following is a description of an improved canvas plant erected and operated in Amador County, by the patentee, Mr. Gates. In this case, the pulp and waste water are conducted from the mill in a flume to the plant, and there divided into two equal streams by the insertion of an adjustable division plate in the flume. The divided pulp passes into boxes (see Fig. 40) 4' long and 1' wide, and having steel screen bottoms with ⅛" and $\frac{1}{16}$" perforations, set on a reversed grade of 6" to the box. The object of these screens is to prevent any chips, leaves, lint, or foreign substance from passing into the sizing-box (Fig. 41) beneath, which consists of a wooden V-shaped trough, 6' long, 15" broad at the top and 2" in the bottom, constructed of 1½" boards. A piece of canvas is tacked on the bottom for packing; underneath is nailed a piece of scantling, 4" x 6", at one end of which, reaching within 2" of the end of the box proper, a slot, 14" long and 2" broad, is cut; here a flattened, galvanized-iron funnel, ending in a 2" pipe, is attached. The pulp falls through the screen with some force and is considerably agitated in the

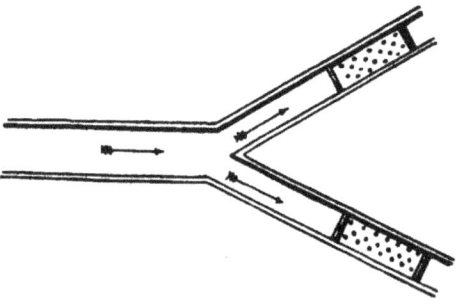

FIG. 40.

separator-box. Naturally the coarser and heavier particles have a tendency to settle toward the bottom. Were the outlet there large enough, all the pulp would pass down and out. Its size of 2" causes the box to fill to the height of a sluice-box in the end, through which the finer pulp flows to the canvas-tables. To facilitate the separation, a device is placed in the lower end, consisting of an iron pipe, ¼" inside diameter, connected with the main pipe above the screen, and divided into two sections, which are connected by rubber hose for ready detachment. The lower 6" of the iron pipe has small perforations, through which clear water is ejected, causing an agitation of the pulp. The end of the pipe is stopped with a wooden plug, easily removed. The agitation at the end of the pipe causes the fine material to be carried upward and into the sluice

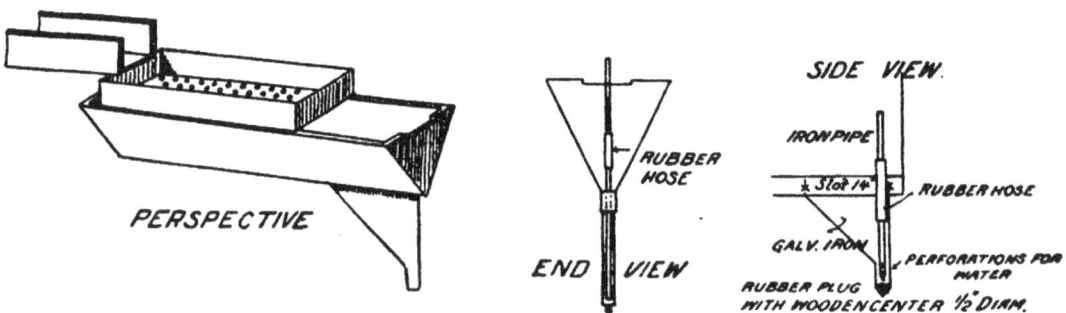

FIG. 41. SIZING-BOX.

at the end of the separator-box. Only coarse sand passes through the bottom pipe, and on examining this with a magnifying glass, very few particles of sulphurets are discernible. This separator works well, and disposes of a lot of coarse, valueless material that would otherwise interfere with the subsequent working of the slimes on the canvas platforms. The fine pulp flowing from the top of the separator is conducted in a sluice to a broad, flat box, in which the stream is divided by partitions into ten separate currents, each terminating over a canvas-table, ten in a row. The pulp goes over a spreader made of strips of galvanized iron, ¾" in height, radiating from a common center to the farthest side of the table, which is 12' wide. These strips are nailed to an inclined board extending across the canvas-table, having an iron strip, 1" high, fastened to the lower end, perforated or notched, with indentations $\frac{1}{16}$" deep and 1" long, affording a perfect distribution. Twenty tables are arranged in two rows of ten each, covered with canvas laid crosswise and overlapping about 2". These tables have a grade of 1½" to the foot, are 13' long and 12' wide. After receiving the flow for an hour, it is shut off from the table and a flow of clear water turned on, which in a few minutes washes away the sand, when it is also stopped; then with a hose ending in a flat nozzle, the accumulated sulphurets are washed from the canvas into a trough below, extending along the base of the entire series. In order to secure sufficient fall for this sluice, each succeeding table is set 4" lower than its predecessor, giving 40" fall on 125' of sluice length. Two extra tables are arranged, one at the end of each row, to take up the surplus flow during the time one of the tables is shut off, to avoid overloading, as each table already carries the proper amount of pulp. The effectiveness of the canvas-tables depends on maintaining an even flow of pulp during a given time; it will not do to overload them. All the pulp that leaves the table is considered waste, and is collected in a flume, to be used a short distance off as power on an overshot wheel, by means of which the patentee runs a vanner of his own invention. This waste water is caught up again and used on a second wheel, which also runs a vanner. The sulphurets washed from the tables flow through a sluice to a box outside the building, 12' long, 2' wide, and 12" deep, with a cross-piece 2' from its upper end, reaching within 2" of the top of the box; in this upper section the coarser grade of the material is retained, while the finer flows over the weir. The two grades are shoveled out separately and placed in separate V-shaped boxes, over which are perforated iron pipes, from which small streams of water trickle, gradually carrying the pulp down and passing it through sluices onto the spreaders of separate vanners. These two machines work with different motions, doing excellent work on this impalpably fine stuff. The slimes flowing from the washing-boxes beneath these vanners are conducted, with the overflow of the two compartment boxes above referred to, to two other canvas-tables, below which they are allowed to escape as waste; not that they have given up all the precious metal they carried, but because the point is reached where it is more economical to lose the remnant than to attempt to save it.

As the slimes from most of the canvas plants, as usually operated (especially where the ore crushed carries a heavy percentage of sulphurets, or has been stamped with a high discharge), are still valuable in gold, they can be conveyed to so-called slime-settlers, or tanks. These tanks, for there are generally several, are placed below the canvas platforms, and are about 2' deep, 2' wide, and 12' to 20' long; they are divided into sections of 2' square, by 2" plank set on edge, extending alternately from each side, leaving an opening 4" wide and 2' deep, causing the slime water to take a serpentine course in passing through. The tanks stand level, and the slimes, in settling, form their own grade as they enter at one end of the tank, and, passing through the successive sections, issue at a diagonally opposite point only slightly clouded. These tanks require cleaning only at long intervals.

Up to the present time, the concentrates in the California mills have been generally handled by the chlorination process, to free them from their gold, but within the last year several plants are successfully working them by the cyanide process.

The tendency in the construction of mills at the present day is to a substitution of steel for iron, where possible, and to an increase in the weight of the stamps.

A greater application of grinding and amalgamating machines, in place of or subsidiary to the stamp-mill, is also noticeable, the most popular of which will be shortly described.

For a more thorough appreciation and knowledge of the work done by mills, records should be kept, by the amalgamator, of all transactions connected with mill work, showing every item, loss of time, consumption of mercury, iron, fuel, water, amount of rock treated, etc., in addition to the records kept in the assay office. This is already being done to some extent, but such records should be kept in the small mills as thoroughly as in the large ones.

GRINDING AND AMALGAMATING MACHINES.

Arrastras.—Although the arrastra has been largely superseded by the stamp-mill, the fact remains that it is the best and cheapest all-round gold-saving appliance we have. Hence, its use is always indicated where small, rich veins are worked in the higher mountain regions, but it is also found valuable placed below the present quartz mill, where the waste waters from the mill can be picked up and used over again for power on horizontal or overshot wheels. In these cases, it handles the tailings from the mill after they have passed over the concentrators and canvas-plants. This part of the milling is usually leased to parties who pay the mine a fixed amount per ton for the tailings, the lessees putting up all their own machinery. These arrastras are built of a size to handle at least 4 tons of tailings in twenty-four hours. Their foundations are either formed of hard rammed clay, concrete, or a plank platform with broken joints, on which a bed of clay is placed. The foundation is always made larger than the circumference of the proposed arrastra. The bed is formed of rocks harder than the substance to be crushed, usually fine-grained basalt, granite, or quartzite. These are picked with a partially level surface, and as near of the same thickness as possible, usually from 1' to 2' thick. They are built around a center cone, forming an annular ring from $2\frac{1}{2}'$ to 6' wide, and are laid with narrow spaces between each rock, into which dry clay should be tightly rammed to within an inch of the surface. The outer circle is formed of rocks or staves, with rammed earth behind, and built from 2' to 4' in height. On the central cone, which consists of stone or a block of wood, and which stands somewhat above the paved bottom, a center post is let in, from which project four arms at right angles to each other, and extending nearly to the outer circle. Heavy, hard rock-drags, weighing from 200 to 1,000 lbs. each (from 400 to 600 lbs. is the usual weight), are attached to the arms by ropes or chains passing through eye-bolts secured in the rock-drags. They are placed so that part of them drag near the cone, with the inside corner slightly in

HORSE-POWER ARRASTRA, KERN COUNTY.

WATER-POWER ARRASTRA, KERN COUNTY.

STEAM-POWER ARRASTRA. KERN COUNTY.

advance, while the remainder sweep near the outer circle with the outer corner in advance. The front edge should always be slightly elevated, so as to permit of the particles passing under the drag instead of being pushed ahead.

Where a horizontal wheel is used, the arms are attached to the center post and the wheel encircles the arrastra, the water striking on buckets set to an angle of 45°. With overshot wheels the arrastra may be run by a belt and pulley attached to the center-post, or by a spur gearing. It requires about 6 H. P. to run an average-sized arrastra. Running tailings, a speed of 15 to 30 revolutions per minute is given; crushing ore, the arrastra should be run slower and the pulp thicker.

For discharging the arrastra, plug-holes at different levels are put into the outer circle, leading the pulp into sluices lined with plates, riffles, and blankets. In some cases the arrastra has been made to work continuously by fitting a screen to a part of the outer circle and letting it discharge into a line of sluices. As the arrastra bottom and drags are extremely uneven and rough when first set up, some coarse sand and water are introduced on first starting, and the drags are allowed to run slowly until somewhat smoothed down, before the regular charge is introduced. The machine is usually only cleaned up thoroughly when the bottom is worn away; between times the crevices are picked out for the depth of an inch or two with picks, scrapers, and spoons, and panned out, with what pulp remains on the bottom, after the charges have been successively thinned down and run off through the plug-holes. If crevicing has been done, a little fresh clay can be rammed in to within 1" of the top of the bed. During the grinding of the charge, the quicksilver is introduced through a cloth; the amalgam should be kept drier than in the stamp battery, though not sufficiently so as to become "crumbly." Great attention must be paid to tamping the bed in solid, otherwise an excessive loss of quicksilver may occur. Continual horn tests of the pulp furnish a guide for the proper working.

Machines have, from time to time, been introduced in California to replace stamps, claiming to do more effective work, both as regards the crushing as well as the amalgamating. Those mostly seen in operation, and finding the most favor, are the Huntington and the Bryan mills, which may be taken as types, and which reduce the ore by a continuous rolling motion; in the one case the roller acting on a ring on the circumference, and in the other on dies in the bottom.

The Huntington Mill consists of a shallow iron pan with a central cone, through which an iron shaft revolves. Bolted on the sides of the pan and inclosing it, are semi-circular iron sections made in two halves and also bolted together; one of these sections contains an opening about 9" deep, divided into three parts, into which curved iron screen-frames are keyed, while the other section contains a feed-trough, attached near the top. Between the bottom of the pan and the lower edge of the screen-frames an iron or steel ring-die fits against the sides of the shallow pan, being secured by wooden wedges; against this die, four rollers, suspended from yokes resting on an iron cover, revolve, receiving their motion from the central shaft. These suspended rollers are pressed by centrifugal force against the ring-die. Each roller is encircled by an iron or steel shoe fastened by wooden wedges; this can be renewed when worn too thin, or when it becomes unround—flattened. Means are provided for lubricating the shafts on which the rollers work, without permitting the lubricant to come in contact with the pulp. As the rollers hang about ½″ above the bottom of the pan, scrapers are attached to the revolving cover between the rollers, and reaching to the bottom of the pan, to prevent the baking of the pulp.

The size of the pan most frequently used is 5′ in diameter, though for prospecting purposes one of 3½′ is also made; the former is run at a speed of 70 revolutions per minute, the latter at 90 revolutions. They are provided with self-feeders, which introduce the ore at regular intervals—the only way in which they can be operated, though not correct in

FIG. 42. THE HUNTINGTON MILL.

principle. A 5′ mill requires about 8 H. P., and crushes about 20 tons per day. Before starting up a certain amount of quicksilver, up to 50 lbs., is introduced into the pan with some water and rock. The supply should be regulated to make a stiffer pulp than in a stamp-battery; quicksilver is added from time to time. A groove in the bottom of the pan, connecting with a plug-hole on the outside, permits of the quicksilver and amalgam being drawn off at intervals to recover the latter, after which the former is returned. If the pan is working correctly the bottom around the center remains bare; this can be observed through the cover while running; when not bare, it is a sign that the pan is being overfed. As the machine throws the pulp with considerable violence through the curved screens, a shield is placed outside of them, directing the pulp into a narrow sluiceway with a spout opening on the apron-plate. It is claimed that the percentage of gold amalgamated and saved on the inside is far greater than in the stamp-mortar, going above 80%; all rusty gold being subjected to a heavy scouring action.

The Russian-iron screens used are short-lived ; they can be made to last somewhat longer by placing a false screen, made from an old worn screen with the openings enlarged, between the pulp and the screen proper.

Great care must be exercised in putting up one of these machines, to get it perfectly level and on a rigid foundation, and to keep all the bolts holding the pan on the foundation well tightened up ; the feed also requires close observation.

When cleaning up or renewing the ring-dies or shoes, the top cover, with the suspended rollers, are lifted out with chain block and tackle, leaving the interior of the pan free for operation.

The mill works well on soft quartz and clayey ores, introduced in pieces not larger than walnuts. A great drawback to the machine is that the rings on the rollers and also the dies become "unround," so that instead of rolling smoothly, they strike in places, necessitating changing the rings before they are worn out; this changing takes up some time.

FIG. 43. THE BRYAN MILL.

The opinions of millmen who have handled the Huntington mill, as to its merits, are very diverse. Where the ore produces a large amount of fine stuff, by using a grizzly with closely set bars, the Huntington can be run to advantage on these "smalls" in conjunction with the stamps.

The Bryan Roller Mill is a modified form of the Chili mill, built in sizes of 4' and 5' diameter. It consists of an annular mortar with an outside gutter and spout, cast solid, containing steel dies arranged in the track of three crushing rollers, which in the 5' mill have a crushing-face of 7", a diameter of 44", and weigh 3,650 lbs. They have fixed axles, "journaled" in a central revolving table, attached to and driven by a belt pulley. This pulley is a cylindrical tank, which, in the smaller pattern, rests immediately on the rollers, and can be made to increase their crushing power by being loaded. The mortar is supplied with curved screen-frames around the entire machine, the pulp being discharged all around into a gutter delivering through a spout, on one side, to an apron-plate.

The chief wearing parts are the steel dies and tires on the rollers; these latter are fastened to the rollers by wooden wedges. According to the statement of the manufacturers (Risdon Iron Works, San Francisco), one set of these wearing parts will crush from 4,000 to 8,000 tons of ore in the large size, and 1,500 to 2,000 tons in the smaller size, and at the rate of 25 to 35 tons and 12 to 20 tons per day, with a speed of 30 and 60 revolutions, respectively, per minute, the smaller size requiring from 5 to 6 H. P. The oil channels for lubricating the bearings are arranged to prevent the oil from entering the mortar. To keep the pulp

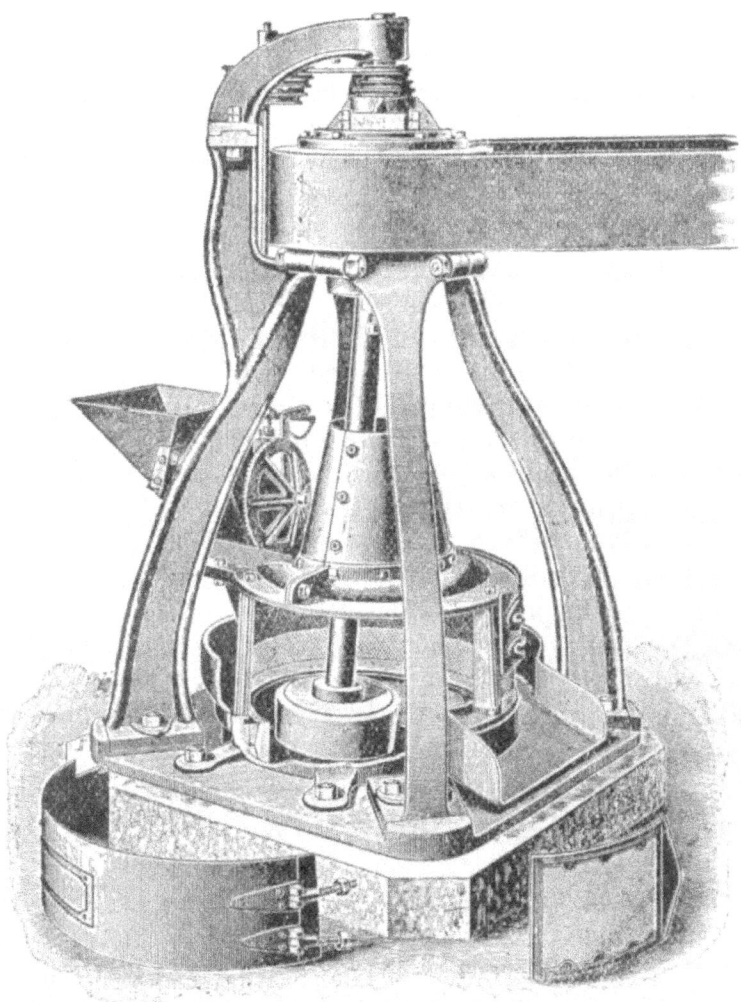

Fig. 44. The Griffin Mill.

from baking to the rollers or dies, and to assist in equalizing the ore received from the feeder, scrapers with adjustable springs follow each roller. They are also provided with self-feeders. In operating the mill, ore, water, and mercury are introduced into the mortar, the pulp passing around next the screens in a current not less than 300' per minute, while the motion inside of the rollers is much slower. The amalgam, working its way toward the center cone, is kept from being re-ground, and can be observed while the mill is in operation; it is claimed to retain 80% of the amalgam in the mortar. To clean it up, the dies between the rollers are removed, the pulp and amalgam taken out, and wooden blocks of the thickness of the die put in their stead, on which the rollers are revolved, when the remaining ones can be taken up. It is claimed for these mills, that they wear smooth, and even while crushing hard quartz, discharge freely (on account of large screen area), avoid sliming and flouring of quicksilver, are good amalgamators, can be cleaned rapidly, are easily put in place, and require small power for amount of work done.

The Griffin Mill belongs to that class of mills using a roll running against a ring or die; but instead of several rollers, as in the Huntington, this has one roller only, swinging from a longer shaft, hung from a point in the central axis of the mill, and rotated about its own axis by the power applied at the top. It is run at a speed of 190 to 200 revolutions per minute, crushing from $1\frac{1}{2}$ to $2\frac{1}{2}$ tons per hour, the power being applied to a horizontal pulley above, from which the shaft is suspended with a universal joint, and the roller is rigidly attached to the lower extremity of the shaft. The roller swings in a circular pan supplied with a ring or die, against which the roller works; and carries on the under side scrapers or plows to prevent the pulp from baking. A circular screen-frame is fastened on the pan, to the top of which a conical shield is attached at the apex, through which the shaft works. The pulley revolves upon a tapered and adjustable bearing, supported by the frame composed of iron standards, two of which are extended above the pulley to carry the arms in which is secured the hollow journal-pin. The shaft is suspended to a universal joint within the pulley. This joint is composed of the ball or sphere with trunnions attached thereto, which work in half boxes that slide up and down recesses in the pulley-head casting. The lubricant is supplied, for all parts needing it, through the hollow pin. The roll revolves within the ring-die in the same direction that the shaft is driven, but on coming in contact with the die, it travels around the die in the opposite direction from that in which the roll is revolving with the shaft. A pressure, by centrifugal force, of 6,000 lbs. is brought to bear on the material being pulverized between the roll and die. The water is introduced with feed when running, and receives a whirling motion from the roll, which brings the pulp against the screens, 9' in area. A circular trough on the outside of the pan conducts the pulp to one side, where it discharges over an apron.

TYPICAL CALIFORNIA GOLD MILLS.

As the details in milling practices of the several counties of the State vary greatly, the following typical mills have been selected to indicate the practice under varying conditions:

No. 1. *Amador County.*—The ore is a soft, easily crushed quartz, with about $1\frac{1}{2}\%$ sulphurets, and is largely mixed with slaty material, which, to the extent of 25%, is found mixed with the concentrates. The stamps weigh 750 lbs. each, and drop 6" about ninety-five times per minute, discharging through a No. 8 slot screen, at the rate of $2\frac{1}{2}$ tons per stamp in twenty-four hours. The stamps drop in the following order: 1, 2, 3, 5, 4; Nos. 1 and 2 having $\frac{1}{4}$" more drop than the other stamps; in the adjoining battery the order is reversed. Iron shoes and dies are used. There is an inside plate used in the battery, which retains about 75% of the amalgam. The apron is 48" x 13", set on a grade of $\frac{3}{4}$" to the foot, and the double sluices below are 9' long by 14" wide, with a grade of $1\frac{1}{2}$" to the foot. From these sluices the pulp passes to vanners. To clean the sulphurets from the slaty admixture, a cradle, 12′ long, 20″ wide, and 4″ deep, has been placed in the mill, run by an eccentric. The dirty slimy sulphurets are taken from the washing-boxes beneath the vanne and placed in a half barrel standing on the floor of the mill, into which a hose is lowered, and the sulphurets

INJECTOR DEVICE
FOR RAISING & CLEANING PULP

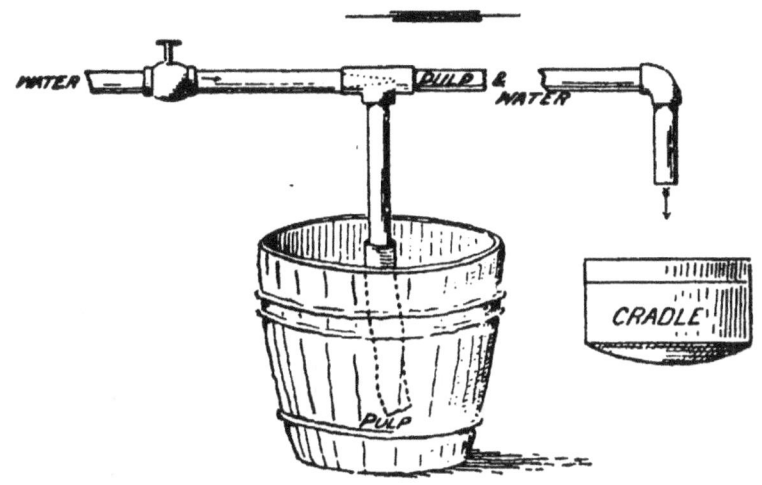

FIG. 45.

are raised from the barrel to the cradle by creating a vacuum, through a small jet of water under pressure forming an ejector. The pulp in the cradle is stirred vigorously toward the head; the grade is from 7″ to 8″ in 12′. This washing in the cradle relieves the pulp of about 25% of waste material. Twelve tons can be washed in a day. The canvas-plant below the vanner has some interesting features. The canvas strips are only 12″ wide. The pulp as it leaves the vanner is carried to a mercury-trap, consisting of a box of diminishing width, with three

QUICKSILVER TRAP.

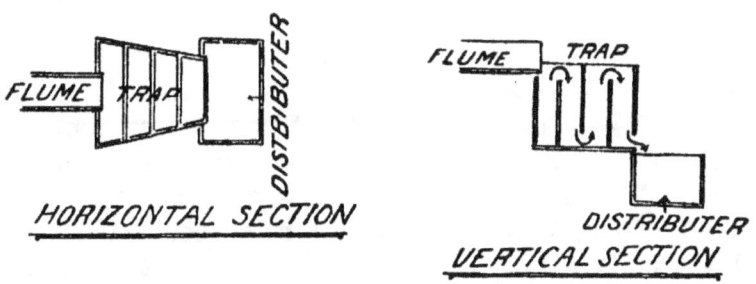

FIG. 46.

upright divisions, under and over which the pulp flows. From the mercury-trap the pulp falls into a long box, about 1′ square at the ends, in the bottom of which are ten holes, whose size is regulated by experience; they must equalize the discharge with the inflow from the mercury-trap. The pulp introduced into the long distributor-box sizes

5—GMP

itself to some extent by gravity; the finer material, being held in suspension longer, finds its way out at the end of the box, while the coarse quickly passes through the holes in the bottom, nearer the center of the box.

There are twelve strips of canvas, 100′ in length, each strip having a width of 12″ and a grade of 4½″ in 12′. The coarse material is all found

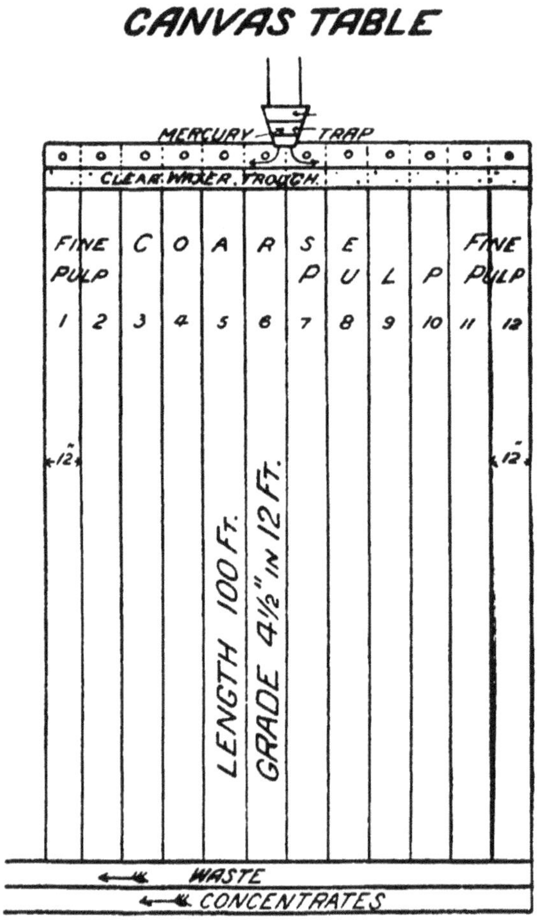

FIG. 47.

on the six center sections, the two outside sections on each side carrying the finer material. An additional series of tables, with 20″ wide sections and a grade of 9″ in 12′, receives the pulp after passing over the first.

No. 2. *Amador County.*—The practice of this mill in handling their tailings may be taken as an example of the better methods now practiced in the State. This mill has 900-lb. stamps, dropping 85 times per minute, with a 6″ drop and a 7″ discharge, kept constant by the use of lower chock-blocks. No. 30 brass-wire screens, 4′ long and 4″ wide, set vertical, are used, giving a duty per stamp of 2½ tons in twenty-four hours. The batteries are supplied with inside front plates. The apron-plates are 46″x30″, set on a grade of 1¾″ to the foot. These are followed by 18′ of sluice-plates, 15″ wide, the first 10′ of which are double. About 66% of the amalgam is recovered in the battery. The loss in quicksilver, which is introduced into the battery every half hour, amounts to about 1½ cents per ton. The total cost of milling at these works is given as 70 cents per ton. The mill is supplied with three vanners to each battery, with $4\frac{1}{2}'$ belts. The pulp from the plate-sluices passes directly to the spreaders of the vanners, a division into thirds being first effected. After leaving the belts, the pulp flows through sluices to a flume, where it is divided into two equal streams by the insertion of an adjustable division plate in the flume. The divided pulp passes to two steel screens with perforations of $\frac{1}{8}''$ and $\frac{1}{16}''$ respectively, which form the bottoms of two 4' boxes, 1' wide, set on a reverse grade of 6" in 4'. These boxes prevent any foreign substance from passing through into the sizing-box below, and clogging the outflow pipes. After the passage of the screens the pulp falls into a separator, consisting of a wooden V-shaped trough, 6' long, 15" wide on top and 2" at the bottom, with a flat, funnel-shaped discharge pipe of galvanized iron attached at one end, ending in a round 2" pipe. As more pulp enters the separator than can be discharged through the 2" pipe, it fills and flows over the end into a launder; the heavier and larger particles sinking down and passing through the pipe. The overflow passes on a spreader that delivers it to a canvas-table, with ten sections; a second similar table, placed below, receives the waste from the first one. The tables are 12' wide, 13' long, and set on a grade of $1\frac{1}{2}''$ to the foot, and to secure a proper grade for the waste-sluice, each section is set 4" below its predecessor. All the waste water passing from the tables is used a short distance off as power on an overshot wheel that runs a vanner, on which are worked the concentrates taken from the tables.

No 3. *Butte County.*—The quartz carries considerable sulphurets. When hoisted from the mine it is dropped immediately over a grizzly, with the bars placed $1\frac{1}{2}''$ apart; the coarse rock crushed is loaded into cars, and trammed to the mill, distant about 150 yards, and dumped into bins which are calculated to carry 1,500 tons. From here chutes convey the ore to the Challenge self-feeders. These are operated from the center stamp in each battery. The stamps, which are supplied with steel shoes and dies, weigh 850 lbs., drop 7", and about 100 times per minute; the discharge is 7"; the screen is No. 8 diagonal-slot, 8" wide; each stamp crushes $2\frac{1}{2}$ tons per twenty-four hours. The screens, which last about four weeks, are used later in the chlorination works for the recovery of cement copper. From the mortar the pulp passes over a 14" mortar plate; thence to a 4' apron and 12' of sluice-plates; aprons and plates are set to a grade of 3" to the foot. The pulp then passes over the vanners, two for each battery, after leaving which, it is conveyed to the canvas-platform house. The canvas-platform is 24' wide and 60' long, covered with x 2 0 0 canvas, and below it are 150' of settling-boxes. The plates are scraped every day, and dressed besides, when required.

No. 4. *Calaveras County.*—The rock consists of massive quartz, schistose and slaty diabase, and chloritic and talcose schist, with iron sulphurets; it is crushed in jaw-breakers at the head of the shaft, after passing over grizzlies, and is dropped into bins, from which the ore is conveyed, in cars, to three other bins in the mill, one for each section of twenty stamps, having a capacity of 600 tons each. The bins discharge into Challenge self-feeders. The sixty stamps weigh 775 lbs. each, and drop 105 times per minute, the drop being 6", and the discharge 10" from the new die. Only one chock-block is used, causing the height of discharge to constantly increase. The duty of the stamps is 4 tons in twenty-four hours. Round-punched tin screens, 10" x 14", are used. They are lightly burned before using. Three and a half of the screen sheets are tacked on the screen-frame on three sides; the top side is secured by a long, narrow strip of wood screwed to the frame. The superficial area of the discharge is about 287 sq. in. The screen-frame is braced by six cross-ribs, to which the screens are tacked. A splash-board is suspended in front of the screen by eyebolts and hooks, with a strip of canvas tacked along the bottom, the full width of the screen. An iron apron or table is secured to the front of the mortar below the screen, the bottom of which falls 1" below the lip of the mortar, permitting the insertion of a rough inch board, 9" in width, in front of the mortar, flush with the upper edge of the lip of the mortar; on this the

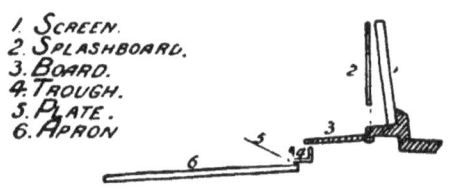

PLATES IN FRONT OF MORTAR

1. SCREEN.
2. SPLASHBOARD.
3. BOARD.
4. TROUGH.
5. PLATE.
6. APRON

FIG. 48.

pulp falls from the screen, and it is claimed to be superior to a plate in retaining the amalgam. Three inches below the board, runs a trough, in which are two apertures one third the distance from each end, which allows the pulp to fall on a short, 6" wide copper plate with a pitch toward the mortar, and from thence to the apron-plate, 2' wide and 24' long, set to a grade of 2" to the foot. An inside front plate is used in the mortar. From the apron-plate the pulp passes to a sluice-box and is conducted to the spreaders of the vanners, of which there are twenty-four. After leaving these, the pulp is led through a sluice-box and flume one mile long to the canvas-plant. The plates are dressed every morning; a battery is hung up, the water shut off, the splash-board removed and washed off, as also the screen and the entire front of the battery, to remove all sand; the plate is then vigorously scoured with a whisk-broom to loosen the amalgam. A very dilute solution of cyanide of potassium is sometimes used during this operation, and the loosened amalgam brushed to the foot of the plate. The plate is then scraped upward with a piece of rubber 4" x 4" and ½" thick; a piece of rubber belting would answer the same purpose. The collected amalgam at the head of the plate is removed in a scoop and placed in a safe. The plate is then sprinkled lightly with quicksilver, which is spread evenly over the entire plate, the water turned on, and the stamps dropped. The operation for all the plates requires nearly three hours.

A clean-up of the mill is made monthly or semi-monthly, according to the condition of the battery amalgam, at which time all necessary repairs are made, and new shoes and dies are fixed if required. Shortly before hanging up, the feed is shut off to permit the ore to be crushed down as low as possible. The water is then shut off from the battery, splash-board and screen removed, and all hosed off; the inside plate is removed, and the amalgam scraped off ' The contents of the battery are now removed, and placed in the revolving clean-up barrel, the dies replaced, tappets set, screens replaced, and the mill started. The pulp .that leaves the mill carries considerable auriferous pyritical slimes, as might be inferred from the high discharge used in the battery. This is conveyed through a sluice 12″ x 8″, to a canvas-plant one mile distant. Just before entering the building set apart for the recovery of these slimes, the sluice is widened to 18″, and divided into three sections by two narrow strips fastened to the bottom. These divisions fork off into separate sluices, which are again subdivided. Three of these subdivis-

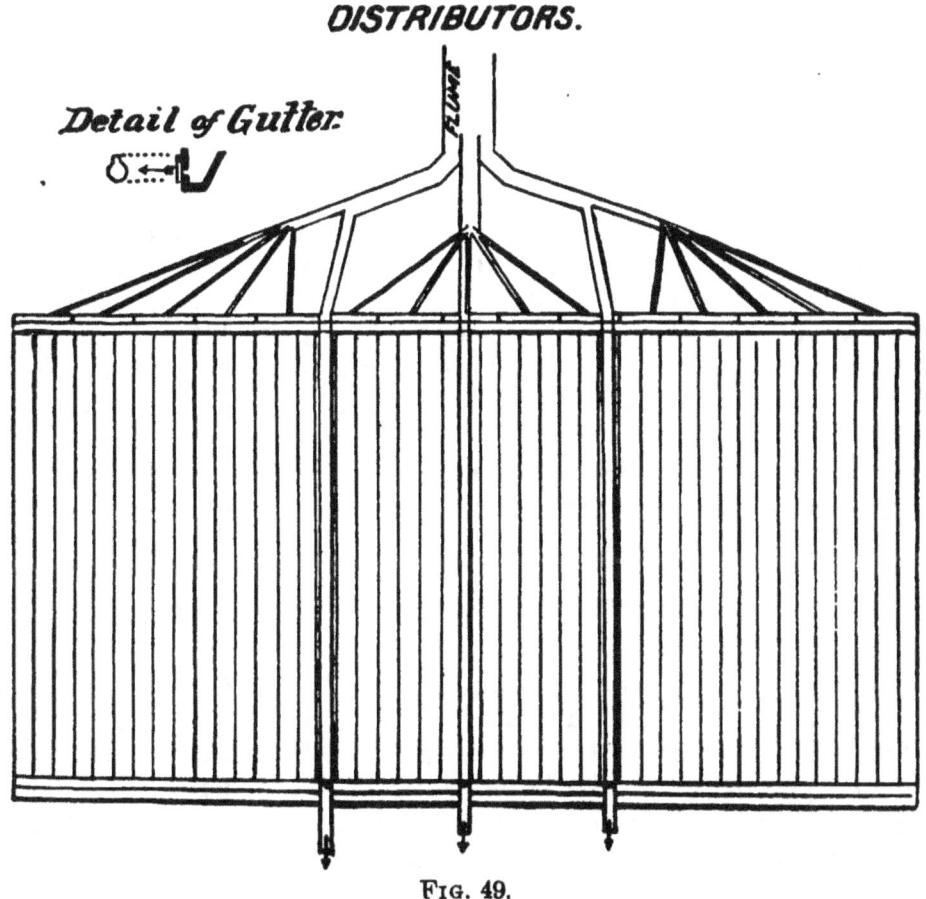

DISTRIBUTORS.

Detail of Gutter.

FIG. 49.

ions are carried directly through the building, and there divided, and the other three are divided off in five separate sluices, one for each section of the canvas-table. There are forty-five sections for each table— ninety in all. They are 42′ long, 22″ wide, and set on a grade $1\frac{1}{8}$″ to the foot. No. 8 duck canvas is used, and when worn on one side it is turned; it lasts about one year. The last division of the pulp, outside the building, is into five boxes, 4″ square, each of which terminates in a receiving-box, reaching across three canvas sections, about $5\frac{1}{2}$′. The five divisions supply one third of the sluices on one side of the building; the pulp passing to the canvas through an auger-hole in the side of the box. The flow is regulated by a slide suspended over the whole. Above the pulp-distributing box is a clear-water box, and at the lower end of the canvas-tables are two sluices, side by side—one to receive the concentrates, the other for the reception and discharge of pulp. The current must be thinned and distributed so that no accumulations form.

No. 5. *El Dorado County.*—The quartz carries considerable slate mixed with it, and about 3% of iron sulphurets. The stamps weigh 950 lbs. each, and drop 4″ 104 times per minute, discharging through a No. 2 sheet-tin, perforated screen with from 5″ to 7″ discharge, crush-. ing 3 tons per stamp in twenty-four hours. The shoes and dies are both steel. The battery is supplied with an inside plate in front. The apron-plates are 16′ long, set on a grade of 1¾″ to the foot, and are followed by 6′ of sluice-plates, 2′ wide, on the same grade. These plates are dressed every day, but only scraped once a month. (Note:—This is not advantageous, as the constant scouring action of the pulp undoubtedly detaches fine particles of amalgam.) The batteries yield 62% of the amalgam. Eight Woodbury concentrators receive the pulp. The quick-silver is introduced into the battery every half hour; the loss of quick-silver being estimated about one pound to every ten tons of ore.

No. 6. *El Dorado County.*—The quartz carries about 2% of sulphurets and contains slate mixed with it. The stamps weigh 750 lbs. each and are fed by Challenge feeders; no rock-breaker is used. The stamps make 96 drops per minute, varying from 4½″ to 6″, with 7″ discharge. The mortars are wide, and have an 8″ wide inside amalgamated plate; the screen is perforated tin, equal to No. 7. The apron is 54″ by 42″, with grade of 1¾″ to the foot, followed by sluice-plates 12′ long, which are double on one battery and divided into four divisions on the other; this latter arrangement gives better results. Below the sluice-plates is a blanket-sluice, 6′ long and 14″ wide, the blanket being washed twice a shift. From these the pulp passes to the Frue vanners with 6′ belts; the same wheel runs both stamps and vanners. The plates are dressed twice in twenty-four hours, but are not scraped until the clean-up, once a month. One and three quarters tons are crushed to the stamp per twenty-four hours.

No. 7. *Mariposa County.*—This mill is working on ores containing gold in a very finely divided state. There are ten stamps of 900 lbs. each, fed by self-feeders. These stamps drop 96 times per minute, with a 7″ discharge while the die is new; when it is worn down one half, a smaller chock-block is placed under the screen. The pulp is retained in the mortar for a long time. The stamps are only raised to the level of the water as it stands in the mortar; and a front inside plate is used. The screen is a 60-mesh. It is claimed that 70% of all the amalgam saved is taken from the inside battery-plate. There are three apron-plates to each battery; the first is immediately in front of the splash-board, next to the lip of the mortar, 12″ deep and the width of the mortar. This is followed by a 2″ drop onto a second plate 3′ deep, across the width of the mortar, succeeded by a 1″ drop to a 4′ apron-plate. From this plate the pulp passes immediately through distributing pipes to the vanners, of which there are three. Two of these are 4′ wide, taking the pulp from one battery, while a 6′ belt vanner takes the pulp from the other battery. The narrow belt vanner gives the best satisfaction.

No. 8. *Nevada County.*—The ore is delivered by car at the top of the mill into grizzlies, the bars of which are 2¼″ apart, and which deliver the coarse stuff to a crusher of the Blake type, through a bin with chute immediately over the crusher, keeping the same constantly supplied without the aid of a shovel. From the rock-breaker the ore is delivered into bins with chutes connecting with the Challenge self-feeders. The feed is operated from the center stamp. The stamps drop 7″ and 86 times per minute, with a 4″ discharge. Steel shoes on iron dies are used. The screens are English sheet-tin, perforated, No. 10, five pieces making a complete screen, costing 50 cents, and lasting one month. The steel shoes on iron dies have records of over 300 tons to the stamp, the daily average being from 1¾ to 2 tons per stamp. The plates are divided into an upper apron, 18″ wide, followed by a 4′ apron. Between the two, catching the pulp from No. 1, is a box 3″ wide, with a perforated screen bottom, somewhat coarser than the battery-screen. This acts as a distributor on No. 2 apron, retaining any coarse pieces. For the forty stamps and accompanying concentrators, 15 miner's inches of water are used, all applied on the inside of the battery. The aprons have a ⅜″ grade to the foot. Below the apron are 12′ of sluice-plates, part of them 30″ wide, while the others have the same width as the apron above. These plates are cleaned up every morning with pieces of rubber belting; it takes about fifteen minutes to clean one set. From the sluice-plates the pulp passes over a 12′ shaking-table covered with silver plates; these plates receive their motion from an eccentric placed underneath. Passing the table the pulp enters a box, from which it is conveyed through pipes to the vanners on a lower floor, two for each battery.

No. 9. *Nevada County.*—The stamps weigh 750 lbs. each, drop 6″, 95 times per minute, with from 4″ to 6″ discharge, and crush from 1¾ to 2 tons per day, using steel shoes and dies. There is a 5″ wide silvered plate in the front of the battery. A No. 9, perforated, sheet-tin screen is used; it is not burnt before putting on, and is turned when the lower edge is worn, lasting on an average 30 days. In front of the screen is a splash-board, provided with an 8″ plate next to the screen. The upper apron-plate is 18″ deep, set on a grade of ¾″ to the foot, with 12′ of apron-plates below, divided into four plates of 3′ each, set on a grade of 1″ to the foot. From these the pulp drops into a box running across the end of the plate, from whence it passes to the vanner. The plates are scraped every twenty-four hours, with the exception of the upper 4″ next the mortar, and are dressed twice a day, using dilute cyanide of potassium. Both rubbers and chisels are used in scraping the plates. In cleaning up the batteries, which occurs once per month, the headings are put into a revolving barrel with pieces of iron and quicksilver, and after running several hours, the contents are removed in buckets, the sand "boiled out" with the hose, the dross skimmed off, and the quicksilver strained. About 75% of the amalgam is saved in the battery. The tailings assay from 25 cts. to $1 50 per ton.

No. 10. *Nevada County.*—The stamps weigh 800 lbs. each, and are given a 6″ drop, 100 times per minute, with a discharge varying from 2″ to 4″; there are no plates in the battery. The ore passes over grizzlies, with bars 1½″ apart, to a No. 2 Blake crusher; thence to the ore-bin that supplies the Challenge feeders, which are operated from the center stamp. Steel shoes and iron dies are used—the shoes lasting, on an average, 155 days; the iron dies, 70 days. No. 6 Russian-iron slot-screens are used. The outside mortar-plate is 14″ wide, with ½″ pitch to the foot, and retains 75% of the plate amalgam. The apron below is 4′ x 4′, with a grade of ¾″ to the foot. Beyond this are 12′ of double sluice-plates 12″ wide, and with ¾″ grade to the foot. Three sand-boxes separate the different apron-plates. From the sluice-plates the pulp passes directly to the concentrators. The duty of the stamps is two tons per day. The tailings assay $1 80 to $2 per ton. From 10 to 12 lbs. of quicksilver per month is used for the 40 stamps. The plates are scraped every day, and the batteries cleaned once a month, the headings being worked in a Knox pan. A weak solution of cyanide of potassium is used in dressing the plates.

No. 11. *Placer County.*—The quartz carries but a small percentage of sulphurets, and is delivered from the mine over an incline tramway to two grizzlies with 12 bars, 3″ apart, 12′ long, 3″ deep, and ½″ wide, set on an angle of 45°. In front, below, and between the grizzlies is a Blake crusher, from which the ore drops into the bin that supplies the Challenge feeders. These are operated from the center stamp. The stamps weigh 750 lbs. each, and drop 5″, 90 times a minute, and the discharge averages 5″. The screen is set on a 4″ block, with a 5″ plate on the inside. The screen is a No. 10, slot-punched, set with a slight incline. Part of the water for the battery is supplied from a small wooden trough, pierced with holes in front of the screen. The outside iron lip of the mortar is covered with a silvered plate. The apron, set on a grade of 1½″ to the foot, is 4′ long, and is followed by 12′ of sluice-plates, 18″ wide. After passing through a quicksilver trap, the pulp passes through a 3″ pipe to the Frue vanners. A tank of quicksilver is used every three months, in crushing 3,500 tons of ore. The plates are scraped every day with rubbers, and are occasionally dressed with phosphate of lime, or with lye. The battery is cleaned out once a week, and yields 50% of the amalgam.

No. 12. *Plumas County.*—The ore is free-milling, and contains about 1¼% of sulphurets. It is delivered to the Blake crushers in the mill by an incline tramway, and the ore passes through the bins to the Challenge feeders. The stamps weigh 850 lbs. each, dropping 8½″, 80 times per minute. The discharge varies from 6″ to 8″, through No. 8 diagonal-slot punched screens, with a discharging surface to each battery of 45″ in length by 6″ in height. The mortar is furnished with a lip plate and a cast-iron trough, which receives the pulp, also with a 5″ inside plate. The pulp passes from the trough to the apron and sluice-plates, which have a grade of 1¾″ to the foot and a length of 20′, and is then passed to the concentrators. Below the mill the tailings are picked up by outside parties and re-ground in arrastras. The tailings assay $2 per ton. The loss of quicksilver at this mill is about a flask for every 4,600 tons crushed. The cost of milling does not exceed 50 cents per ton when using water-power. The plates are cleaned every twenty-four hours. About 60% of the amalgam is derived from the batteries, which are cleaned up once a month. The headings are placed in an iron revolving barrel, and the panning-out is done with a batea.

No. 13. *Plumas County.*—The ore is hauled to the mill by wagon, and is broken and fed by hand. The stamps weigh 750 lbs. each, drop 5″ to 6″, 80 times per minute, with a discharge varying from 6″ to 8″, through a No. 9 slot-punched, Russian-iron screen, crushing 1¾ tons per stamp per twenty-four hours. The battery is supplied with an inside plate, about 6″ wide, attached to the screen; the latter is set slightly inclined. The screen-frame leaves about 4″ at the upper end of the mortar-front open, in front of which and reaching nearly to the lip is a canvas curtain. The apron-plate is 5′ x 4½′, set on a grade of 1″ in 11″; below the apron is a drop-box, from which the pulp passes to the sluice-plates; these are 10′ long by 15″ wide. The aprons are scraped every day with rubbber belting, and the plate on the screen is cleaned once or twice a week. In dressing the plates, brine with an addition of sulphuric acid is used. About 20% of the amalgam is saved in the batteries, and about 80% on the plates. Neither concentrators nor canvas-tables are used. One tank of quicksilver is used every six months, using twenty stamps.

No. 14. *Shasta County.*—The ores carries 1½% of iron and copper sulphurets, besides free gold, averaging $9 per ton. There are 30 stamps, weighing 850 lbs. each, supplied with Challenge feeders, working from the second stamp. These stamps are hung and dropped somewhat at variance with the usual custom, No. 1, the end stamp on the left, being placed 1″ farther from the side than is No. 5, the end stamp on the right, and the sequence of the drop is 5, 4, 3, 1, 2; the stamps never rising out of the water. It is claimed that by this arrangement a better swash is obtained in the battery. The stamps drop 5″, 92 times per minute, with a discharge of 6″ to 7″, and crushing 2 tons per stamp per twenty-four hours. The mortar is supplied with front and back inside plates. The apron-plate is 4′ x 4′, set on a grade of 1¼″ to the foot, followed by a double set of sluice-plates, 16″ wide and 16′ long, with a grade of 1″ to the foot. The apron-plate is kept rather wet with mercury by frequent dressing. Burr-slot screens, Nos. 40 and 45, are used. About 66% of the amalgam is derived from the battery. The pulp is concentrated on four Triumph and ten Frue vanners, and is then passed to two canvas-platforms, 36′ and 24′ long, respectively, and 20′ wide, divided into sections 2½′ in width, covered with twill instead of canvas, which is said to give equally good results, and is considerably cheaper. These tables have a grade of 1¼″ to the foot. The plates are scraped once a day, and the mill is cleaned up twice a month. The company chlorinate their own sulphurets, roasting in a small two-hearth furnace, with a capacity of one ton per twenty-four hours.

No. 15. *Sierra County.*—On account of topography, the ore has to be elevated by a lift to the top of the mill. The stamps weigh 850 lbs. each, and drop 5″, 80 times per minute, with a 6″ discharge through No. 7 slot-cut screens. The cams, bosses, and tappets are steel; the shoes and dies iron. The apron is 4′x4′, with a grade of 1¾″ to the foot, and is followed by a double sluice-plate, 16″ wide, 12′ long, and pitched 1½″ to the foot. The plates are not scraped at regular intervals; in dressing them, lye is used occasionally. The plate on the screen, 6″x52″, is cleaned every other day. About 86% of the amalgam is saved in the battery, and the tailings only show a trace of gold. Johnston concentrators receive the pulp from the plates; these concentrators are run with 110 to 112 side-strokes per minute, the belt revolving once in seven minutes. The waste from the concentrators is forced to a higher level by an "ejector," and then passes through a pointed box. The heavy material is then passed through a series of drop-boxes and discharged into the river.

No. 16. *Sierra County.*—The ore carries a considerable amount of clay, and is delivered to the mill over an incline track to a Blake crusher. The stamps weigh 850 lbs. each, and make 75 to 78 drops of 6" per minute, with a discharge varying from 7" to 9", using a No. 10 slot-cut screen. The inside of the mortar is furnished with front and back plates, the former 8", the latter 4" wide. Cast-iron shoes and dies are used, crushing 1½ tons to the stamp per day. The order in which the stamps drop is 1, 5, 2, 4, 3. The apron-plate is the width of the mortar, is 6' long, and is set to a grade of 2½" to the foot, followed by 12' of sluice-plates, 14" wide. As there are but few sulphurets, no concentrators are in the mill. The apron and sluices are dressed every day, but only scraped once a month; cyanide of potassium is used in dressing the plates. The battery is cleaned once a month. About 5 lbs. of quicksilver is used to every 1,500 tons. About 70% of the amalgam is obtained from the battery.

No. 17. *Tuolumne County.*—This mill of 10 stamps crushes quartz containing little or no free gold, but with 3 per cent of sulphurets, chiefly iron. The stamps weigh 1,000 lbs. each (fed by self-feeders), working at 96 drops of 6", crushing 2½ tons per stamp. Chrome steel shoes and dies are used, which wear about 1" per week. No. 30 brass wire screens are used, the screen having a slight inclination, 10°. There are no plates used on the inside of the battery, and only one apron-plate, 4½'x 6', to each five stamps, which is dressed daily. It is set to a grade of 1¾" to the foot. Nearly all of the amalgam is derived from this apron. The pulp passes from the aprons through a series of troughs to four Frue concentrators with corrugated belts, using a large amount of water. These are succeeded by wooden troughs, 8" wide at top, spreading to 16", which divide into three troughs, carrying equal amounts of pulp, thinned by adding 3 miner's inches of clear water above the forks. These deliver into a V-trough running at the head of a canvas-platform, divided into twelve sections, 22" wide and 75' long, set on a grade of 1½" to the foot. These tables are covered with No. 7 duck, which lasts 90 days. The V-trough has extending over its entire length a square 8" trough, for clear water. The pulp flows from the V-trough through 1" auger-holes (two to each section of the canvas), supplied with wooden plugs to regulate the flow of the pulp, the clear-water trough being similarly supplied. Every half hour the flow of the pulp is arrested on a section, while the flow of clear water is continued until the lighter sands are washed off, leaving the sulphurets on the canvas. The flow from the tables is delivered into V-boxes running across the end of the canvas-platform, with a grade from the outer edges to the center, and delivered to a second canvas-table trough, where it undergoes a similar treatment. After removing the lighter sands from the upper platform, a sheet-iron pan is placed below the end, which extends to a separate trough, into which the sulphurets adhering to the canvas are swept with the aid of the flowing clear water, and conveyed to a settling-tank divided into two sections in such a way that when one section is filled, the sulphurets are run into the second section, allowing the first to be shoveled out, each section being treated consecutively. When No. 1 is being swept, No. 2 has the pulp-flow shut off, making the operation continuous. After sweeping down a section, the plugs from the pulp trough are removed and the clear water shut off, permitting the concentration to be renewed. It requires the attention of two persons, night and day, to attend to these two platforms. Everything but the sweepings pass over the second canvas-platform, and then go to waste. The sweepings from the second table are treated in a manner similar to those

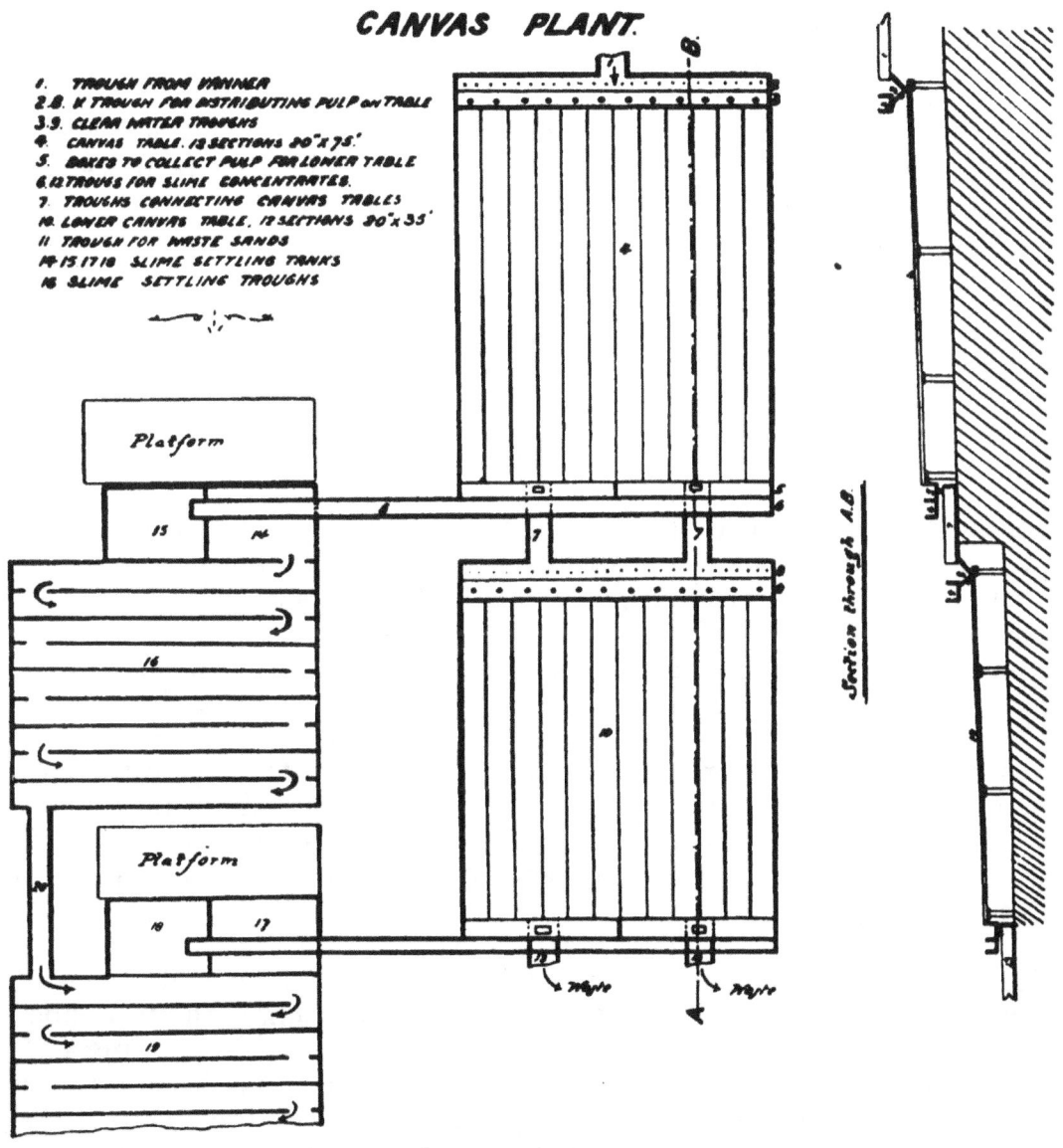

Fig. 50.

from the upper one. Accessory to these two sulphuret settling-tanks are two large slime settling-tanks; these are divided into twenty sections, 2' square, 20' long, divided from each other by 2" plank extending entirely across the tank. Within 2' of the ends, on alternate sides of these divisions, slots 4" deep and 2' wide are cut to permit the water, which is clouded with the fine slimes, to pass from one division to the other. At the end farthest from the entrance, the water (still somewhat clouded) runs to waste. The slime water from the first settling-tank passes through the second. The sulphuret settling-tanks are shoveled out every few days, while the slime settling-tanks remain undisturbed for months. The material from the vanners, upper and lower settling-tanks, and sometimes from the slimes, is mixed by weight before going to the chlorination works. There are 2,400 sq. ft. of canvas-tables and 5,000 sq. ft. of settling-floor.

No. 18. *Tuolumne County.*—This mill is working on quartz carrying both free gold and sulphurets. The ore is delivered from the mine at the top of the mill, into a general ore-bin, for the entire 40 stamps, after passing through the rock-breaker. The bin has a capacity of 650 tons, and delivers the ore into the self-feeders direct. The stamps weigh 850 lbs. each, and steel shoes and dies are used. Each stamp has its separate guide, made of two blocks of hard maple, fitted together and bored through to receive the stem. The front block is first put in place, the stem set in, and the rear block dropped in behind a cast-iron piece,

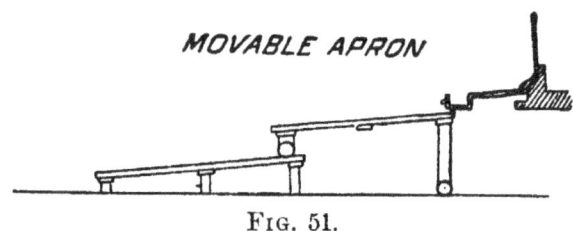

MOVABLE APRON

FIG. 51.

which is secured by wooden wedges driven in from above, so that when required it can easily be removed. The pulp from the battery falls over a 9″ silvered plate, the width of the mortar, into a box 12″ square, supplied with six 1″ holes, 8″ apart, near the front lower edge, that permit the pulp to flow onto a 4′x 5′ silvered plate, divided in two parts by a wooden strip down the center. The fall from the box to the plate is $3\frac{1}{2}$″. The apron-plate is mounted on a carriage, which can be pushed back, giving access to the battery, the 4″ grooved wheels in front running on a half-round iron strip placed on the sides of the lower plate-frame. From the movable apron the pulp passes over 12′ of plates, divided into three 4′ sections, with a dividing strip down the center. Sixteen concentrators are used.

No. 19. *Tuolumne County.*—This mill presents some peculiarities in its construction. There are 10 stamps of 850 lbs. each, with steel shoes and dies. The stamps are given 100 drops per minute, dropping $4\frac{1}{2}$″, with only $2\frac{1}{2}$″ discharge to commence on, through a No. 50 brass-wire screen. The rock-breaker (Wheeler pattern) and ore-bin are set on a rock foundation, the frames being entirely disconnected from the rest of the building, to counteract the vibratory motion of the crusher. In placing the mortar-block and mortar, the space around the block was filled in with concrete, and a double thickness of tanned belting laid between the block and the mortar, after which a fire was built in the latter until it settled into the belting. The wear of the shoes and dies is about 1″ per month, and the duty of the stamps $1\frac{1}{4}$ tons per stamp. When the dies are partially worn, a 2″ iron plate is placed under them, to maintain a regular discharge. The inside of each mortar is provided with cast-iron side plates and a sheet-iron covered board at the back, to prevent wear on the mortar. The apron-plates are set 12″ out from the mortar. The pulp from the battery passes over a 9″ plate into double-pointed boxes of iron, bolted on the front of the mortar, and thence through a couple of 2″ pipes to a spreader and to the silver-plated apron. The apron is followed by double sluice-plates, each 2′ wide and 10′ long, all set on a grade of 1¼″ to the foot. Two thirds of the amalgam is obtained from the battery. No concentrators are used.

SPECIFICATIONS FOR A FORTY-STAMP GOLD MILL (WATER POWER).*

MACHINERY.

Water Wheels and Pulleys.—One water wheel, 6′ in diameter, to drive the battery; the wheel to be supplied with a shaft, boxes, collars, gate, and nozzle, automatic governor, and a pulley 36″ in diameter, grooved for 1½″ manilla ropes.

One driving pulley, 12′ in diameter.

One idler pulley, 48″ inches in diameter, grooved for one 1½″ rope, and fitted with shaft and boxes.

One slack-tightener pulley, 48″ in diameter, grooved for one 1½″ rope, and fitted with shaft, boxes, carriage, track, and counterbalance weight.

The rope for transmission is to be put on in one piece, passing around the idler and slack-tightener (which are to be set on an angle in such a way that they will take the rope from one side of one of the main pulleys and pass it on to the opposite side of the other pulley), thereby making but one splice in the whole rope, which will be kept in constant tension by the slack-tightener.

One wheel, 4′ in diameter, to drive the rock-breakers; the wheel to be supplied with a shaft, boxes, collars, gate, and nozzle, and a pulley 34″ in diameter, grooved for one 1½″ manilla rope.

One driving pulley, 60″ in diameter.

One idler pulley, 30″ in diameter, grooved for one 1½″ rope, and fitted with shaft and boxes.

One slack-tightener pulley, 30″ in diameter, grooved for one 1½″ rope, and fitted with shaft, boxes, carriage, track, and counterbalance weight; rope to be put on similar to that for the battery.

One wheel, 36″ in diameter, to drive the concentrators; the wheel to be supplied with shaft, boxes, collars, gate, and nozzle, automatic governor, and a pulley 16″ in diameter, grooved for one 1″ manilla rope.

One driving pulley, 48″ in diameter.

Forty-Stamp Battery; stamps to weigh 850 to 900 lbs. each, arranged to run in eight batteries of five stamps each, by belts and friction clutch pulleys from battery line shaft.

Eight high cast-iron *Mortars*, single discharge, each to weigh about 5,000 lbs.; to be planed all over the bottom, and faced where the apron joins on; eight holes to be accurately cored in the base for 1½″ anchor bolts. Each mortar to have five cast-iron linings. The aggregate weight of these linings is about 500 lbs. per mortar.

Eight cast-iron *Aprons*, to be faced where they join on to the mortars, and fastened in place with ¾″ bolts.

---

* From the VIIIth Report of State Mineralogist, 1888, p. 728.

Eight sugar pine *Screen Frames*, to have iron facings put on the ends where the keys bear against them; the edges to be fitted with dowel-pins to join them to the inside plate-block.

Sixteen inside *Plate-Blocks*, two sets, one to be 6" high, and the other to be 4" high; to be well fitted into the mortars, and to have plates fitted and fastened on with brass screws; blocks to be bolted together to keep them from splitting, and to be fitted with iron facings where the keys bear against them, and well fitted to the screw frames.

Eight brass wire *Screens*, No. 30 mesh, to be fastened on to the screw frames with copper tacks.

Sixteen gilt-headed *End Keys*, for screen frames, to be well fitted in place.

Sixteen *Bottom Keys*, for screen frames, to be well fitted in place.

Sixty-four *Foundation Bolts*, for mortars, to be $1\frac{1}{2}$" in diameter by 30" long, with hexagon nuts on the top ends and steel keys in the bottom ends.

Sixty-four wrought-iron *Washers*, $4"x4"x\frac{3}{4}"$, for bottom ends of foundation bolts.

Eight sheets of *Rubber*, $\frac{1}{4}"$ thick by 30" by 60", for mortar foundation. Mill blankets tarred may be used in place of rubber.

Forty chrome steel or cast-iron *Dies*, 9" in diameter by 7" high, with square base well fitted into the mortars, 10" from center to center.

Forty chrome steel *Shoes*, 9" in diameter by 8" high, with tapered shank $3\frac{3}{4}"$ in diameter at top end, $4\frac{3}{4}"$ in diameter at bottom end, by 5" long, to fit into the stamp-heads by being covered with dry, hard pine $\frac{3}{8}"$ thick; this being driven in by being allowed to drop a few times on the bare die.

Forty chrome steel *Stamp-Heads*, 9" in diameter by 17" long, with a conical socket cored into the lower end, 4" in diameter at inner end and $5\frac{1}{8}"$ in diameter at the outer end, and $5\frac{1}{2}"$ deep, and a conical socket cored and actually bored out to fit the tapered end of the stamp stem, $2\frac{1}{4}"$ in diameter at inner end, and $3\frac{1}{4}"$ gauge at the outer end, by 6" deep. Transverse rectangular keyways are to be cored through the stem-head, $1"x2\frac{1}{2}"$, for loosening the shoes and stems.

Two steel *Loosening Keys*, $\frac{1}{4}"$ thick by 1" at the point (2" at the head) by 18" long, for loosening the shoes and stems.

Forty best refined iron or mild steel *Stems*, turned perfectly true, full length, $3\frac{1}{4}"$ gauge by 14' long, to be tapered on both ends to accurately fit the stamp-heads. Each stem weighs about 360 lbs.

Forty chrome steel, double-faced *Tappets*, 9" in diameter by 11" long, with a steel gib and two steel keys accurately fitted in place; both faces to be turned true; tappets to be bored with the gibs in place to accurately fit the stems, and to be counter-bored opposite the gibs by moving the center $\frac{1}{4}"$ away and, with diameter $\frac{1}{8}"$ less than the bore, taking a cut $\frac{1}{8}"$ deep. Each tappet weighs 112 lbs.

Eight *Upper* and eight *Lower Guides*, with cast-iron frames; guide-blocks to be made of good, dry maple timber and well fitted in place; the guides may also be made entirely of wood.

Four extra quality, mild steel *Cam-Shafts*, turned true full length, 5½" gauge diameter by 14' long; key-seated for cams and pulley; key-seats must not run through the bearings.

Ten heavy *Corner Boxes*, 5½" gauge bore; eight of them to be 12" long, and two to be 20" long; all of them to be planed all over the bottoms and backs, and furnished with bolts 1" in diameter, to fasten them to the battery frame.

Forty double-armed, chrome steel *Cams*—twenty right and twenty left hand—to be made 29" long over all, the hub to be 11" in diameter and 5½" through the bore; the lifting faces to be 2½" wide, and ground smooth; the hubs to be bored to fit the shaft accurately, and properly key-seated and fitted with steel keys, and each marked to their respective places, giving them a combination as follows: Counting from the left-hand side, when facing the battery, throughout the full ten stamps of each cam-shaft, No. 1 cam will drop its stamp first; then Nos. 8, 4, 10, 2, 7, 5, 9, 3, and 6 consecutively. This is the order: 1, 4, 2, 5, 3. Each cam weighs about 158 lbs. The curve of the face of the cam is the involute of a circle, usually slightly modified.

Four pairs of cast-iron double *Sleeve Flanges*, for wood pulleys; flanges to be 36" in diameter, and 14" through the bore; to be turned all over the inside, where they fit on the wood; the outside flange is to be bored and fitted to the sleeve and fastened with a gib-headed steel key; the hub to be bored and fitted to the cam-shaft and fastened with a steel key.

Four *Wood Pulleys*, 72" in diameter by 17" face; to be made of best kiln-dried sugar pine, and all joints to be filled with white lead in oil; the cast-iron flanges to be well fitted on and bolted with twelve ⅞" bolts.

Eight wrought-iron *Collars*, for cam-shaft, 5½" bore, fitted with two steel set-screws in each.

Eight wrought-iron *Jack-Shafts*, 3" in diameter by 60" long; black finish.

Sixteen cast-iron *Jack-Shaft Side Brackets*, with four lag-screws, ⅝" by 6", for each, to fasten them in place.

Forty open *Latch Sockets*, lined with leather.

Forty wood *Finger-bars*, to be fitted and bolted to the above sockets, and furnished with wrought-iron caps and handles.

A complete set of *Water-Pipes* for a battery of forty stamps, with all fittings, cocks, and connections.

Bolts and Washers for Battery Frame.—Six brace rods, 1¼"x25', 7" between two nuts; 6 brace rods, 1¼"x12', 6" between two nuts; 26 bolts for mudsills, 1"x30"; 24 bolts for yokes, 1"x28"; 24 bolts for yokes, 1"x52"; 48 bolts for guide girts, 1"x32"; 4 bolts for knee beam, 1"x28"; 36 splice bolts for mudsills, ⅞", 16" between head and nut; 12 splice bolts for tail girt, ⅝", 9½" between head and nut; 32 bolts for mortar-blocks, 1", 59" from point to point; 64 bolts for mortar-blocks, 1", 65" from point to point; 24 joint bolts for posts, 1", 35" between two nuts; 6 joint bolts for knee posts, 1", 45" between two nuts; 6 joint bolts for knee posts, 1", 35" between two nuts; 24 joint bolts for knee beams, 1",

43″ between two nuts; 10 joint bolts for tail girts, 1″, 21″ between two nuts; 24 cast-iron washers for 1¼″ rods; 514 cast-iron washers for 1″ bolts; 72 cast-iron washers for ⅞″ bolts; 24 cast-iron washers for ⅝″ bolts; 40 sheet-iron washers, 3½″ square by ¼″ thick, for 1″ joint bolts.

Battery Line Shafting and Pulleys.—One shaft, 5½″ gauge by 18′ long, properly key-seated; one shaft, 5″ gauge by 15′ 6″ long, properly key-seated; one shaft, 5″ gauge by 17′ long, properly key-seated; one shaft, 4″ gauge by 17′ long, properly key-seated; two shafts, 3″ gauge by 10′ 6″ long, properly key-seated; two face couplings, 5″ gauge, properly fitted and keyed in place; one face coupling, 4″ gauge, properly fitted and keyed in place; two face couplings, 3″ gauge, properly fitted and keyed in place; two babbitted boxes, 5½″ gauge; three babbitted boxes, 5″ gauge; two babbitted boxes, 4″ gauge; two babbitted boxes, 3″ gauge; all of the above boxes to be made of the same height, planed all over the bottoms, with drip cups cast on to the sides, and furnished with suitable bolts to fasten them to the 16″ battery knee beams; two collars for 5¼″ shafting, with two steel set-screws in each; four friction clutch pulleys, 48″ in diameter and 16½″ face, complete, with levers and connections; pulleys to be fitted to line shaft in their proper places, with phosphor-bronze bushings, the drivers to be properly keyed on with steel keys; one pulley, 6′ in diameter, grooved for three 1½″ manilla ropes, pulley to be well balanced and keyed to the shaft with a steel key.

Water-Pipes.—Sufficient 3″ pipes and fittings to connect battery pipes with feed-water tanks.

Traveling Hoist.—One traveling crab, with track-iron and rails, to extend full length of battery.

One 2-ton Weston's differential chain-block.

Ore-Feeders.—Eight Challenge self-feeders, complete, for batteries, with all connections.

Ore-Bin Gates.—Eight ore-bin gates, 18″x 24″, for fine ore, with guides, racks, pinions, shafts, boxes, hand-wheels, and bolts.

Three ore-bin gates, 24″x 36″, for coarse ore, with guides, racks, pinions, shafts, boxes, hand-wheels, and bolts.

Sluices and Aprons.—Eight cast-iron aprons, 54″ wide by 56″ long, to be fitted under the lip of the mortar apron.

Eight silver-plated copper plates, 54″x 56″x ⅛″, to be made of best Lake Superior copper, and to have one ounce of silver per square foot; plates to be fitted into the cast-iron aprons, and fastened by strips of wood on the sides, which are bolted to the sides of the apron.

Eight cast-iron sluices, 54″ wide by 12′ long, to be made into two sections and bolted together by flanges, the lower section to have a quicksilver trap or trough cast on to the end, extending the full width of the sluice, and to have a connection made for a 2″ pipe to conduct the pulp to the dividing tanks, and thence to the concentrators.

Twenty-four silver-plated copper plates, 54″x 48″x ⅛″, to be made of best Lake Superior copper, and to have one ounce of silver per square foot; plates to be fitted into the sluices, overlapping at the joints, and to be fastened in place in the same manner as those in the apron.

There are to be eight silver-plated copper shaking-tables, one for each battery, placed below the apron-plates. These tables consist of a light iron framework suspended upon movable springs. This table is given a longitudinal oscillation by means of eccentrics.

Dividing Tanks and Pulp Pipes.—Eight cast-iron dividing tanks, 10" long by 8" wide by 6" deep, with 2" pipe connection in one end and two $1\frac{1}{2}$" pipe connections in the other end, each to have a wooden swinging tongue put in so as to direct the pulp to either of the $1\frac{1}{2}$" pipes, or a part to the one and a part to the other. The tanks are to be connected with the sluices by 2" pipes, and with the concentrators by $1\frac{1}{2}$" pipes.

Inside Plates and Blocks.—Three wooden blocks for each mortar, to be 3", $4\frac{1}{2}$", and 6" high, respectively, to be fitted into the mortars under the screw frames; each block to have iron facings, fitted in flush and screwed on where the keys come, and to have a silver-plated copper plate bent to the proper shape and screwed on with silver-plated brass screws; the copper plates to be made of best Lake Superior copper, 6"x 50"x $\frac{3}{16}$", and to have one ounce of silver per square foot.

Concentrators and Shafting.—Sixteen endless belt concentrators, complete, with water-pipes and fittings to connect with supply tanks. All sulphuret tanks, complete, to be made of good redwood lumber.

One piece of shafting, $2\frac{1}{2}$"x 16'; six pieces of shafting, 2"x 16'; three pieces of shafting, 2"x 10'; eight face couplings, 2"; four babbitted boxes, $2\frac{1}{2}$", with bolts for 8' timber; eighteen babbitted boxes, 2", with bolts for 8" timber; two collars, $2\frac{1}{2}$", with steel set-screws; two collars, 2", with steel set-screws; one pulley, 48" in diameter, grooved for one 1" rope, and properly fitted and keyed with a steel key to $2\frac{1}{2}$" shaft; two pulleys, 6" face by 36" in diameter, properly fitted and keyed with steel keys to the 2" shaft; sixteen pulleys, 4" face and 10" in diameter, properly fitted and keyed with steel keys to the 2" shaft; sixteen loose pulleys, 4" face by 10" in diameter, properly fitted to the 2" shaft; sixteen collars, with steel set-screws for same.

Rock-Breakers and Shafting.—Two rock-breakers, 9"x 15"; one piece shafting, 4"x 16'; one piece shafting, $3\frac{1}{2}$"x 16'; one piece shafting, 3"x 16'; one face coupling, $3\frac{1}{2}$"; one face coupling, 3"; three babbitted boxes, 4", with bolts for 10" timber; two babbitted boxes, $3\frac{1}{2}$", with bolts for 10" timber; two babbitted boxes, 3", with bolts for 10" timber; two collars, 4", with steel set-screws; one pulley, 48" in diameter, grooved for 1" and $1\frac{1}{2}$" manilla rope, and properly fitted and keyed to the 4" shaft, with a steel key; three pulleys, 20" straight face by 20" in diameter, properly fitted and keyed to the shafting.

Clean-up Barrel.—One clean-up barrel, 24" inside diameter by 48" inside length, to be made of cast-iron $1\frac{1}{2}$" thick, with two discharge openings, $5\frac{1}{2}$" in diameter, in the sides diametrically opposite each other, the heads and discharge doors to be accurately fitted; journals to be 4" gauge, cast on to the heads; one tight and one loose pulley, 7" face by 30" in diameter; two babbitted boxes, 4" gauge; one driving pulley, 6" in diameter by 14" face.

Batea.—One batea, 48" in diameter, with gears and hangers complete, and tight and loose pulleys, $4\frac{1}{2}$" face by 16" in diameter; one driving pulley, 9" face by 21" in diameter.

Machinery for Clean-up Room.—One clean-up pan, 24" inside diameter, with tight and loose pulleys.

One driving pulley, 8″ face by 16″ in diameter.

One cast-iron washing-tank, 24″ by 30″ by 24″ deep, with three pipe connections for drawing off water.

One cast-iron washing-tank, 30″ by 36″ by 24″ deep, with three pipe connections for drawing off water.

One cast-iron washing-tank, 30″ by 54″ by 30″ deep, with three pipe connections for drawing off water.

One marble top, complete, for washing-tanks.

One side washstand, with pipes and fittings.

All pipes and fittings necessary to bring water to the clean-up pan and washing-tanks.

Retort and Melting Furnace.—One retort, 10″x 36″, inside dimensions, with amalgam trays, condenser, catch tank, furnace front, bearers, bars, smokestack, and base plate, guy rods, dampers, binders, and all pipes and fittings to bring water to the condenser.

One cast-iron melting furnace, complete, with doors, grate-bars, bearers, cast-iron shell, and damper.

Two bullion molds for 500 and 750 ounces.

Four black-lead crucibles, No. 16, with covers.

One crucible tongs for No. 16 crucible.

One skimmer for bullion.

Transmission Ropes and Belts.—Six hundred feet best manilla or cotton rope, 1½″ in diameter, to drive battery line shaft.

Two hundred and fifty feet best manilla rope, 1½″ in diameter, to drive rock-breaker line shaft.

One hundred and fifty feet best manilla rope, 1″ in diameter, to drive concentrator line shaft.

Two hundred feet best rubber belting, 16″ by 5-ply, for batteries.

One hundred and eighty feet best rubber belting, 10″ by 4-ply, for rock-breakers.

Thirty-two feet best rubber belting, 7″ by 4-ply, for clean-up barrel.

Sixty-five feet best rubber belting, 6″ by 4-ply, for batea.

Thirty feet best rubber belting, 6″ by 4-ply, for concentrator shafting.

Four hundred and twenty feet best rubber belting, 3″ by 4-ply, for concentrators.

Thirty feet best rubber belting, 3″ by 4-ply, for clean-up pan.

BUILDINGS, AND ERECTION OF MILL, ETC.

Stonework.—All foundations and retaining walls to be built of large stone, properly banded, and well laid in cement mortar, composed of ten parts good, clear sand, two parts good quality of lime, and one part best Portland cement, special care being taken to keep all dirt or clayey material excluded; all exposed faces of retaining walls to be well pointed up and finished with the same material.

Ore-Bins.—Mudsills to be made of 12″x14″ timbers, laid flatwise; foundation posts to be made of 14″x14″ timbers; sills, posts, and caps for ore-bins proper to be made of 12″x12″ timbers, the posts to be boxed 1″ into the sills and caps; braces for incline bottom, to be made of 10″x 12″ timbers; supporting braces to be made of 8″x 12″ timbers. All planking to be 3″ thick and lined throughout with 1″ boards, to break joints over the planks.

Battery Frame.—Mudsills to be made of 14"x16" sugar pine, or good yellow pine free from sap; to be well bedded in concrete, which must be put on the clean bedrock. Linesills to be made of 12"x16" and 20"x16" sugar pine or yellow pine, of good quality, to be well bolted down to the mudsills.

Mortar-blocks to be made of two pieces each, to be 30" thick and wide enough to fill space between the linesills and battery posts; all to be sized and well fitted. The timbers for mortar-blocks are to be accurately fitted together and secured with six 1" bolts, and two oak keys, 4" wide by 5" thick at the point and 6" at the head. Keys to be accurately fitted and firmly driven. Blocks to be sized and finished above the floors.

Yokes to be made of 10"x10" timber, well fitted and bolted to the linesills and battery posts.

Battery posts to be made of 12"x24" and 20"x24", good quality pine timber, to be dressed all over, and bolted down to the linesills with 1" joint bolts, the large posts to be made with double tennon on the bottom. The knee beams to be made of 12"x16" timber, dressed all over. The knee posts to be made of 12"x16" timber, dressed all over. The stringer on top of the knee posts to be made of 12"x16" timber in two pieces, to be spliced with a ship splice 3' long, stringer to be dressed all over. Knee posts to be framed into stringer with double tennons; outside stringer at end of knee beams to be made of 8"x12" timber in two pieces, spliced with ship splices in center 3' long, and to be dressed all over.

Bottom guide girt to be made of 12"x16" timber, dressed all over, one piece for each twenty-stamp battery, and to extend past the outside posts 12"; the top girt to be made of 12"x14" timber, dressed all over, and made the same length as the lower ones; all braces to be made of 8"x12" timber, dressed all over, and framed with double tennons; no keys are to be used in braces or guide girts, but they must be accurately fitted without.

All boxing about battery frame to be ½" deep, and where braces or knee beams are smaller than the timbers they frame into, they must be housed in ½" deep; *i. e.*, the timber must not be boxed out clear across.

The cam-shaft is to be set 4¾" from the center to the center of the stems.

A 2" plank floor is to be put on top of the knee beams, which is to be planed on the under side; also, a 2" double board floor to be put in back of the battery, on about the same level as the knee beams.

The whole battery frame to be painted with two coats of light-cream paint, properly mixed with oil, and the wood pulleys and guides to be painted blue, the iron work to be painted black. The out-board bearing frame to be made of 12"x16" timber, planed all over, well framed and bolted together, and anchored to a solid stone foundation, as shown in plan, and to be painted same as battery frame.

Water Wheel Frames are to be made of 12"x12" lumber throughout, well anchored down to a stone foundation. That part of the frame which comes above the floor is to be dressed and painted the same as the battery frame.

The water wheels are to be housed with tongued and grooved lumber, 4" wide.

BUILDINGS.

Frame Work.—Ore-house main frame is to be made of 8″x8″ timbers throughout, with 3″x6″ girts and studding.

Battery and concentrator rooms frame is to be made of 8″x10″ posts and chords, 6″x10″ sills, 8″x8″ principal rafters and straining beams, 4″x8″ truss braces, and 3″x6″ girts and studding.

Clean-up, sulphuret, and water wheel rooms main frames are to be made of 8″x8″ timbers, with 3″x6″ girts and studding.

Floors.—Ore-house floors to be made of one thickness of 2″ planks.

Battery, concentrator, and water wheel rooms floors are to made of 1″x 8″ lumber, double thickness, surfaced on top, to be supported on 3″x 6″ joists 18″ apart.

Sulphuret and clean-up rooms floors are to be made of concrete laid on top of a heavy wood floor, which is to be supported on foundations made of 8″x 8″ timbers.

Roofs.—All roofs are to be made with 2″x 8″ rafters 18″ apart, with 1″x 6″ board 4″ apart, and covered with No. 26 standing seam, painted, iron roofing.

Walls.—All walls are to be covered with 1″x 10″ rustic.

Cornices.—All cornices are to project 24″, measured horizontally from the walls of building, with a 12″ frieze and a 5″ facia made of dressed lumber.

Windows.—All windows, except those for sulphuret room, are to be made of twelve lights of 10″x 16″ glass, and frames made to suit of dressed lumber, with casing outside 5″ wide.

Twelve windows are to be put in the ore-house, seven windows in the battery room, six windows in the clean-up room, twelve windows in the sulphuret room, and five windows in the water wheel room.

Skylights.—Six skylights, made of twelve lights of 10″x 20″ glass, to be put into the roof of the concentrator room.

Doors.—All doors, both sliding and swinging, to be 3′x 7′x 1¾″ thick, with panels.

Two sliding doors are to be put in the ore-house, and one outside swinging door in the battery room; one swinging door leading from the battery room to the clean-up room; two sliding doors leading from the concentrator room to the sulphuret room; two outside sliding doors for the sulphuret room, and one outside swinging door for the water wheel room.

All doors to be set in good substantial casings, outside cased with surfaced lumber, and furnished with all trimmings and locks.

Stairs.—There is to be a flight of stairs at each end of the mill, one flight leading from the battery room floor to the floors above, and one flight of stairs from the battery room floor to the concentrator room floor.

All stair stringers to be made of 2″x 12″ lumber, and treads of 2″x 10″ lumber.

Hand Rails are to be put on to the outside of all stairs and around the landings of same, also in front of the battery room floor and all other floors and platforms where there is danger of falling. All to be made of dressed lumber, well painted.

Retort House and Assay Office, to be 20' wide by 48' long, with a retort and melting furnace room, a weighing room, and a storeroom; the two latter to be lath and plaster finished, and the whole building to be finished similar to the mill buildings, with iron roof, rustic, etc.

Paint and Whitewash.—All buildings are to be painted on the outside with a good coat of brown mineral paint, and the window and door casings and cornices to be painted with two coats of white lead paint.

The mill to be whitewashed throughout the inside, including the building frame, ore-bins, etc.

Tanks.—There are to be two 4,000-gallon redwood tanks, 3" stock, set up at the end of the mill upon strong timber foundations, and one tank 8' wide by 10' long by 4' high, inside measurements, to be made of 3" planks, with 8" x 8" frame; planking to be well fitted together, and properly caulked inside with oakum. The latter tank is to be set at the end of the last sluice-box coming out of the mill.

Drain Boxes and Tailings Sluices.—Battery sluices and aprons to be set on framework so arranged that the grade can be changed easily. This framework to be planed all over. Sluices and frames to be painted same as battery frame.

There will be a sluice in front of battery room floor, made of surfaced lumber; also to be painted and so arranged as to conduct any water away which drips from the floor.

There will be sluices put in under the concentrator room floor, two of which will be 6" wide by 8" deep, to run lengthwise to catch the tailings from the concentrators, and one to be 8" wide by 10" deep, to run cross-wise and to take the tailings from the first two sluices, and conduct the same outside. All tailings sluices to have a fall of one in twelve, and to be made of 2" lumber, well fitted and nailed together. Proper sluices from the clean-up room, to conduct water and tailings therefrom, must be connected to tailings sluices under concentrator room.

All sulphuret boxes, and drain boxes for concentrators, to be made of good quality of redwood lumber, 1¼" thick, dressed on both sides, and well fitted and screwed together.

The weight of all parts is 240,000 lbs., and there are 325,000 feet of lumber in the building.

Specifications for a canvas plant are not considered necessary, as the construction is extremely simple and no standard has been adopted. Full descriptions are given in the preceding pages.

GOLD RUSH BOOKS

OREGON, USA

www.GoldMiningBooks.com

Books On Mining

Visit: www.goldminingbooks.com to order your copies or ask your favorite book seller to offer them.

Mining Books by Kerby Jackson

Gold Dust: Stories From Oregon's Mining Years - Oregon mining historian and prospector, Kerby Jackson, brings you a treasure trove of seventeen stories on Southern Oregon's rich history of gold prospecting, the prospectors and their discoveries, and the breathtaking areas they settled in and made homes. 5" X 8", 98 ppgs. Retail Price: $11.99

The Golden Trail: More Stories From Oregon's Mining Years - In his follow-up to "Gold Dust: Stories of Oregon's Mining Years", this time around, Jackson brings us twelve tales from Oregon's Gold Rush, including the story about the first gold strike on Canyon Creek in Grant County, about the old timers who found gold by the pail full at the Victor Mine near Galice, how Iradel Bray discovered a rich ledge of gold on the Coquille River during the height of the Rogue River War, a tale of two elderly miners on the hunt for a lost mine in the Cascade Mountains, details about the discovery of the famous Armstrong Nugget and others. 5" X 8", 70 ppgs. Retail Price: $10.99

Oregon Mining Books

Geology and Mineral Resources of Josephine County, Oregon - Unavailable since the 1970's, this important publication was originally compiled by the Oregon Department of Geology and Mineral Industries and includes important details on the economic geology and mineral resources of this important mining area in South Western Oregon. Included are notes on the history, geology and development of important mines, as well as insights into the mining of gold, copper, nickel, limestone, chromium and other minerals found in large quantities in Josephine County, Oregon. 8.5" X 11", 54 ppgs. Retail Price: $9.99

Mines and Prospects of the Mount Reuben Mining District - Unavailable since 1947, this important publication was originally compiled by geologist Elton Youngberg of the Oregon Department of Geology and Mineral Industries and includes detailed descriptions, histories and the geology of the Mount Reuben Mining District in Josephine County, Oregon. Included are notes on the history, geology, development and assay statistics, as well as underground maps of all the major mines and prospects in the vicinity of this much neglected mining district. 8.5" X 11", 48 ppgs. Retail Price: $9.99

The Granite Mining District - Notes on the history, geology and development of important mines in the well known Granite Mining District which is located in Grant County, Oregon. Some of the mines discussed include the Ajax, Blue Ribbon, Buffalo, Continental, Cougar-Independence, Magnolia, New York, Standard and the Tillicum. Also included are many rare maps pertaining to the mines in the area. 8.5" X 11", 48 ppgs. Retail Price: $9.99

Ore Deposits of the Takilma and Waldo Mining Districts of Josephine County, Oregon - The Waldo and Takilma mining districts are most notable for the fact that the earliest large scale mining of placer gold and copper in Oregon took place in these two areas. Included are details about some of the earliest large gold mines in the state such as the Llano de Oro, High Gravel, Cameron, Platerica, Deep Gravel and others, as well as copper mines such as the famous Queen of Bronze mine, the Waldo, Lily and Cowboy mines. This volume also includes six maps and 20 original illustrations. 8.5" X 11", 74 ppgs. Retail Price: $9.99

Metal Mines of Douglas, Coos and Curry Counties, Oregon - Oregon mining historian Kerby Jackson introduces us to a classic work on Oregon's mining history in this important re-issue of Bulletin 14C Volume 1, otherwise known as the Douglas, Coos & Curry Counties, Oregon Metal Mines Handbook. Unavailable since 1940, this important publication was originally compiled by the Oregon Department of Geology and Mineral Industries includes detailed descriptions, histories and the geology of over 250 metallic mineral mines and prospects in this rugged area of South West Oregon. 8.5" X 11", 158 ppgs. Retail Price: $19.99

Metal Mines of Jackson County, Oregon - Unavailable since 1943, this important publication was originally compiled by the Oregon Department of Geology and Mineral Industries includes detailed descriptions, histories and the geology of over 450 metallic mineral mines and prospects in Jackson County, Oregon. Included are such famous gold mining areas as Gold Hill, Jacksonville, Sterling and the Upper Applegate. 8.5" X 11", 220 ppgs. **Retail Price: $24.99**

Metal Mines of Josephine County, Oregon - Oregon mining historian Kerby Jackson introduces us to a classic work on Oregon's mining history in this important re-issue of Bulletin 14C, otherwise known as the Josephine County, Oregon Metal Mines Handbook. Unavailable since 1952, this important publication was originally compiled by the Oregon Department of Geology and Mineral Industries includes detailed descriptions, histories and the geology of over 500 metallic mineral mines and prospects in Josephine County, Oregon. 8.5" X 11", 250 ppgs. **Retail Price: $24.99**

Metal Mines of North East Oregon - Oregon mining historian Kerby Jackson introduces us to a classic work on Oregon's mining history in this important re-issue of Bulletin 14A and 14B, otherwise known as the North East Oregon Metal Mines Handbook. Unavailable since 1941, this important publication was originally compiled by the Oregon Department of Geology and Mineral Industries and includes detailed descriptions, histories and the geology of over 750 metallic mineral mines and prospects in North Eastern Oregon. 8.5" X 11", 310 ppgs. **Retail Price: $29.99**

Metal Mines of North West Oregon - Oregon mining historian Kerby Jackson introduces us to a classic work on Oregon's mining history in this important re-issue of Bulletin 14D, otherwise known as the North West Oregon Metal Mines Handbook. Unavailable since 1951, this important publication was originally compiled by the Oregon Department of Geology and Mineral Industries and includes detailed descriptions, histories and the geology of over 250 metallic mineral mines and prospects in North Western Oregon. 8.5" X 11", 182 ppgs. **Retail Price: $19.99**

Mines and Prospects of Oregon - Mining historian Kerby Jackson introduces us to a classic mining work by the Oregon Bureau of Mines in this important re-issue of The Handbook of Mines and Prospects of Oregon. Unavailable since 1916, this publication includes important insights into hundreds of gold, silver, copper, coal, limestone and other mines that operated in the State of Oregon around the turn of the 19th Century. Included are not only geological details on early mines throughout Oregon, but also insights into their history, production, locations and in some cases, also included are rare maps of their underground workings. 8.5" X 11", 314 ppgs. **Retail Price: $24.99**

Lode Gold of the Klamath Mountains of Northern California and South West Oregon
(See California Mining Books)

Mineral Resources of South West Oregon - Unavailable since 1914, this publication includes important insights into dozens of mines that once operated in South West Oregon, including the famous gold fields of Josephine and Jackson Counties, as well as the Coal Mines of Coos County. Included are not only geological details on early mines throughout South West Oregon, but also insights into their history, production and locations. 8.5" X 11", 154 ppgs. **Retail Price: $11.99**

Chromite Mining in The Klamath Mountains of California and Oregon
(See California Mining Books)

Southern Oregon Mineral Wealth - Unavailable since 1904, this rare publication provides a unique snapshot into the mines that were operating in the area at the time. Included are not only geological details on early mines throughout South West Oregon, but also insights into their history, production and locations. Some of the mining areas include Grave Creek, Greenback, Wolf Creek, Jump Off Joe Creek, Granite Hill, Galice, Mount Reuben, Gold Hill, Galls Creek, Kane Creek, Sardine Creek, Birdseye Creek, Evans Creek, Foots Creek, Jacksonville, Ashland, the Applegate River, Waldo, Kerby and the Illinois River, Althouse and Sucker Creek, as well as insights into local copper mining and other topics. 8.5" X 11", 64 ppgs. **Retail Price: $8.99**

Geology and Ore Deposits of the Takilma and Waldo Mining Districts - Unavailable since the 1933, this publication was originally compiled by the United States Geological Survey and includes details on gold and copper mining in the Takilma and Waldo Districts of Josephine County, Oregon. The Waldo and Takilma mining districts are most notable for the fact that the earliest large scale mining of placer gold and copper in Oregon took place in these two areas. Included in this report are details about some of the earliest large gold mines in the state such as the Llano de Oro, High Gravel, Cameron, Platerica, Deep Gravel and others, as well as copper mines such as the famous Queen of Bronze mine, the Waldo, Lily and Cowboy mines. In addition to geological examinations, insights are also provided into the production, day to day operations and early histories of these mines, as well as calculations of known mineral reserves in the area. This volume also includes six maps and 20 original illustrations. 8.5" X 11", 74 ppgs. **Retail Price: $9.99**

Gold Mines of Oregon - Oregon mining historian Kerby Jackson introduces us to a classic work on Oregon's mining history in this important re-issue of Bulletin 61, otherwise known as "Gold and Silver In Oregon". Unavailable since 1968, this important publication was originally compiled by geologists Howard C. Brooks and Len Ramp of the Oregon Department of Geology and Mineral Industries and includes detailed descriptions, histories and the geology of over 450 gold mines Oregon. Included are notes on the history, geology and gold production statistics of all the major mining areas in Oregon including the Klamath Mountains, the Blue Mountains and the North Cascades. While gold is where you find it, as every miner knows, the path to success is to prospect for gold where it was previously found. **8.5″ X 11″, 344 ppgs. Retail Price: $24.99**

Mines and Mineral Resources of Curry County Oregon - Originally published in 1916, this important publication on Oregon Mining has not been available for nearly a century. Included are rare insights into the history, production and locations of dozens of gold mines in Curry County, Oregon, as well as detailed information on important Oregon mining districts in that area such as those at Agness, Bald Face Creek, Mule Creek, Boulder Creek, China Diggings, Collier Creek, Elk River, Gold Beach, Rock Creek, Sixes River and elsewhere. Particular attention is especially paid to the famous beach gold deposits of this portion of the Oregon Coast. **8.5″ X 11″, 140 ppgs. Retail Price: $11.99**

Chromite Mining in South West Oregon - Originally published in 1961, this important publication on Oregon Mining has not been available for nearly a century. Included are rare insights into the history, production and locations of nearly 300 chromite mines in South Western Oregon. **8.5″ X 11″, 184 ppgs. Retail Price: $14.99**

Mineral Resources of Douglas County Oregon - Originally published in 1972, this important publication on Oregon Mining has not been available for nearly forty years. Included are rare insights into the geology, history, production and locations of numerous gold mines and other mining properties in Douglas County, Oregon. **8.5″ X 11″, 124 ppgs. Retail Price: $11.99**

Mineral Resources of Coos County Oregon - Originally published in 1972, this important publication on Oregon Mining has not been available for nearly forty years. Included are rare insights into the geology, history, production and locations of numerous gold mines and other mining properties in Coos County, Oregon. **8.5″ X 11″, 100 ppgs. Retail Price: $11.99**

Mineral Resources of Lane County Oregon - Originally published in 1938, this important publication on Oregon Mining has not been available for nearly seventy five years. Included are extremely rare insights into the geology and mines of Lane County, Oregon, in particular in the Bohemia, Blue River, Oakridge, Black Butte and Winberry Mining Districts. **8.5″ X 11″, 82 ppgs. Retail Price: $9.99**

Mineral Resources of the Upper Chetco River of Oregon: Including the Kalmiopsis Wilderness - Originally published in 1975, this important publication on Oregon Mining has not been available for nearly forty years. Withdrawn under the 1872 Mining Act since 1984, real insight into the minerals resources and mines of the Upper Chetco River has long been unavailable due to the remoteness of the area. Despite this, the decades of battle between property owners and environmental extremists over the last private mining inholding in the area has continued to pique the interest of those interested in mining and other forms of natural resource use. Gold mining began in the area in the 1850's and has a rich history in this geographic area, even if the facts surrounding it are little known. Included are twenty two rare photographs, as well as insights into the Becca and Morning Mine, the Emmly Mine (also known as Emily Camp), the Frazier Mine, the Golden Dream or Higgins Mine, Hustis Mine, Peck Mine and others. **8.5″ X 11″, 64 ppgs. Retail Price: $8.99**

Gold Dredging in Oregon - Originally published in 1939, this important publication on Oregon Mining has not been available for nearly seventy five years. Included are extremely rare insights into the history and day to day operations of the dragline and bucketline gold dredges that once worked the placer gold fields of South West and North East Oregon in decades gone by. Also included are details into the areas that were worked by gold dredges in Josephine, Jackson, Baker and Grant counties, as well as the economic factors that impacted this mining method. This volume also offers a unique look into the values of river bottom land in relation to both farming and mining, in how farm lands were mined, re-soiled and reclamated after the dredges worked them. Featured are hard to find maps of the gold dredge fields, as well as rare photographs from a bygone era. **8.5″ X 11″, 86 ppgs. Retail Price: $8.99**

Quick Silver Mining in Oregon - Originally published in 1963, this important publication on Oregon Mining has not been available for over fifty years. This publication includes details into the history and production of Elemental Mercury or Quicksilver in the State of Oregon. **8.5″ X 11″, 238 ppgs. Retail Price: $15.99**

Mines of the Greenhorn Mining District of Grant County Oregon - Originally published in 1948, this important publication on Oregon Mining has not been available for over sixty five years. In this publication are rare insights into the mines of the famous Greenhorn Mining District of Grant County, Oregon, especially the famous Morning Mine. Also included are details on the Tempest, Tiger, Bi-Metallic, Windsor, Psyche, Big Johnny, Snow Creek, Banzette and Paramount Mines, as well as prospects in the vicinities in the famous mining areas of Mormon Basin, Vinegar Basin and Desolation Creek. Included are hard to find mine maps and dozens of rare photographs from the bygone era of Grant County's rich mining history. **8.5″ X 11″, 72 ppgs. Retail Price: $9.99**

Geology of the Wallowa Mountains of Oregon: Part I (Volume 1) - Originally published in 1938, this important publication on Oregon Mining has not been available for nearly seventy five years. Included are details on the geology of this unique portion of North Eastern Oregon. This is the first part of a two book series on the area. Accompanying the text are rare photographs and historic maps.**8.5" X 11", 92 ppgs. Retail Price: $9.99**

Geology of the Wallowa Mountains of Oregon: Part II (Volume 2) - Originally published in 1938, this important publication on Oregon Mining has not been available for nearly seventy five years. Included are details on the geology of this unique portion of North Eastern Oregon. This is the first part of a two book series on the area. Accompanying the text are rare photographs and historic maps.**8.5" X 11", 94 ppgs. Retail Price: $9.99**

Field Identification of Minerals For Oregon Prospectors - Originally published in 1940, this important publication on Oregon Mining has not been available for nearly seventy five years. Included in this volume is an easy system for testing and identifying a wide range of minerals that might be found by prospectors, geologists and rockhounds in the State of Oregon, as well as in other locales. Topics include how to put together your own field testing kit and how to conduct rudimentary tests in the field. This volume is written in a clear and concise way to make it useful even for beginners. **8.5" X 11", 158 ppgs. Retail Price: $14.99**

The Bohemia Mining District of Oregon - Originally published in 1900, this important publication on Oregon Mining has not been available for over a century. Included in this volume are important insights into the famous Bohemia Mining District of Oregon, including the histories and locations of important gold mines in the area such as the Ophir Mine, Clarence, Acturas, Peek-a-boo, White Swan, Combination Mine, the Musick Mine, The California, White Ghost, The Mystery, Wall Street, Vesuvius, Story, Lizzie Bullock, Delta, Elsie Dora, Golden Slipper, Broadway, Champion Mine, Knott, Noonday, Helena, White Wings, Riverside and others. Also included are notes on the nearby Blue River Mining District. **8.5" X 11", 58 ppgs. Retail Price: $9.99**

The Gold Fields of Eastern Oregon - Unavailable since 1900, this publication was originally compiled by the Baker City Chamber of Commerce Offering important insights into the gold mining history of Eastern Oregon, "The Gold Fields of Eastern Oregon" sheds a rare light on many of the gold mines that were operating at the turn of the 19th Century in Baker County and Grant County in North Eastern Oregon. Some of the areas featured include the Cable Cove District, Baisely-Elhorn, Granite, Red Boy, Bonanza, Susanville, Sparta, Virtue, Vaughn, Sumpter, Burnt River, Rye Valley and other mining districts. Included is basic information on not only many gold mines that are well known to those interested in Eastern Oregon mining history, but also many mines and prospects which have been mostly lost to the passage of time. Accompanying are numerous rare photos **8.5" X 11", 78 ppgs. Retail Price: $10.99**

Gold Mining in Eastern Oregon - Originally published in 1938, this important publication on Oregon Mining has not been available for over a century. Included in this volume are important insights into the famous mining districts of Eastern Oregon during the late 1930's. Particular attention is given to those gold mines with milling and concentrating facilities in the Greenhorn, Red Boy, Alamo, Bonanza, Granite, Cable Cove, Cracker Creek, Virtue, Keating, Medical Springs, Sanger, Sparta, Chicken Creek, Mormon Basin, Connor Creek, Cornucopia and the Bull Run Mining Districts. Some of the mines featured include the Ben Harrison, North Pole-Columbia, Highland Maxwell, Baisley-Elkhorn, White Swan, Balm Creek, Twin Baby, Gem of Sparta, New Deal, Gleason, Gifford-Johnson, Cornucopia, Record, Bull Run, Orion and others. Of particular interest are the mill flow sheets and descriptions of milling operations of these mines. **8.5" X 11", 68 ppgs. Retail Price: $8.99**

The Gold Belt of the Blue Mountains of Oregon - Originally published in 1901, this important publication on Oregon Mining has not been available for over a century. Included in this volume are rare insights into the gold deposits of the Blue Mountains of North East Oregon, including the history of their early discovery and early production. Extensive details are offered on this important mining area's mineralogy and economic geology, as well as insights into nearby gold placers, silver deposits and copper deposits. Featured are the Elkhorn and Rock Creek mining districts, the Pocahontas district, Auburn and Minersville districts, Sumpter and Cracker Creek, Cable Cove, the Camp Carson district, Granite, Alamo, Greenhorn, Robinsonville, the Upper Burnt River Valley and Bonanza districts, Susanville, Quartzburg, Canyon Creek, Virtue, the Copper Butte district, the North Powder River, Sparta, Eagle Creek, Cornucopia, Pine Creek, Lower Powder River, the Upper Snake River Canyon, Rye Valley, Lower Burnt River Valley, Mormon Basin, the Malheur and Clarks Creek districts, Sutton Creek and others. Of particular interest are important details on numerous gold mines and prospects in these mining districts, including their locations, histories, geology and other important information, as well as information on silver, copper and fire opal deposits. **8.5" X 11", 250 ppgs. Retail Price: $24.99**

<u>Mining in the Cascades Range of Oregon</u> - Originally published in 1938, this important publication on Oregon Mining has not been available for over seventy five years. Included in this volume are rare insights into the gold mines and other types of metal mines in the Cascades Mountain Range of Oregon. Some of the important mining areas covered include the famous Bohemia Mining District, the North Santiam Mining District, Quartzville Mining District, Blue River Mining District, Fall Creek Mining District, Oakridge District, Zinc District, Buzzard-Al Sarena District, Grand Cove, Climax District and Barron Mining District. Of particular interest are important details on over 100 mines and prospects in these mining districts, including their locations, histories, geology and other important information. **8.5″ X 11″, 170 ppgs. Retail Price: $14.99**

Idaho Mining Books

<u>Gold in Idaho</u> - Unavailable since the 1940's, this publication was originally compiled by the Idaho Bureau of Mines and includes details on gold mining in Idaho. Included is not only raw data on gold production in Idaho, but also valuable insight into where gold may be found in Idaho, as well as practical information on the gold bearing rocks and other geological features that will assist those looking for placer and lode gold in the State of Idaho. This volume also includes thirteen gold maps that greatly enhance the practical usability of the information contained in this small book detailing where to find gold in Idaho. **8.5″ X 11″, 72 ppgs. Retail Price: $9.99**

<u>Geology of the Couer D'Alene Mining District of Idaho</u> - Unavailable since 1961, this publication was originally compiled by the Idaho Bureau of Mines and Geology and includes details on the mining of gold, silver and other minerals in the famous Coeur D'Alene Mining District in Northern Idaho. Included are details on the early history of the Coeur D'Alene Mining District, local tectonic settings, ore deposit features, information on the mineral belts of the Osburn Fault, as well as detailed information on the famous Bunker Hill Mine, the Dayrock Mine, Galena Mine, Lucky Friday Mine and the infamous Sunshine Mine. This volume also includes sixteen hard to find maps. **8.5″ X 11″, 70 ppgs. Retail Price: $9.99**

<u>The Gold Camps and Silver Cities of Idaho</u> - Originally published in 1963, this important publication on Idaho Mining has not been available for nearly fifty years. Included are rare insights into the history of Idaho's Gold Rush, as well as the mad craze for silver in the Idaho Panhandle. Documented in fine detail are the early mining excitements at Boise Basin, at South Boise, in the Owyhees, at Deadwood, Long Valley, Stanley Basin and Robinson Bar, at Atlanta, on the famous Boise River, Volcano, Little Smokey, Banner, Boise Ridge, Hailey, Leesburg, Lemhi, Pearl, at South Mountain, Shoup and Ulysses, Yellow Jacket and Loon Creek. The story follows with the appearance of Chinese miners at the new mining camps on the Snake River, Black Pine, Yankee Fork, Bay Horse, Clayton, Heath, Seven Devils, Gibbonsville, Vienna and Sawtooth City. Also included are special sections on the Idaho Lead and Silver mines of the late 1800's, as well as the mining discoveries of the early 1900's that paved the way for Idaho's modern mining and mineral industry. Lavishly illustrated with rare historic photos, this volume provides a one of a kind documentary into Idaho's mining history that is sure to be enjoyed by not only modern miners and prospectors who still scour the hills in search of nature's treasures, but also those enjoy history and tromping through overgrown ghost towns and long abandoned mining camps. **8.5″ X 11″, 186 ppgs. Retail Price: $14.99**

<u>Ore Deposits and Mining in North Western Custer County Idaho</u> - Unavailable since 1913, this important publication was originally published by the Us Department of the Interior and has been unavailable for a century. Included are fine details on the geology, geography, gold placers and gold and silver bearing quartz veins of the mining region of North West Custer County, Idaho. Of particular interest is a rare look at the mines and prospects of the region, including those such as the Ramshorn Mine, SkyLark, Riverview, Excelsior, Beardsley, Pacific, Hoosier, Silver Brick, Forest Rose and dozens of others in the Bay Horse Mining District. Also covered are the mines of the Yankee Fork District such as the Lucky Boy, Badger, Black, Enterprise, Charles Dickens, Morrison, Golden Sunbeam, Montana, Golden Gate and others, as well as those in the Loon Mining District. **8.5″ X 11″, 126 ppgs. Retail Price: $12.99**

<u>Gold Rush To Idaho</u> - Unavailable since 1963, this important publication was originally published by the Idaho Bureau of Mines and has been unavailable for 50 years. "Gold Rush To Idaho" revisits the earliest years of the discovery of gold in Idaho Territory and introduces us to the conditions that the pioneer gold seekers met when they blazed a trail through the wilderness of Idaho's mountains and discovered the precious yellow metal at Oro Fino and Pierce. Subsequent rushes followed at places like Elk City, Newsome, Clearwater Station, Florence, Warrens and elsewhere. Of particular interest is a rare look at the hardships that the first miners in Idaho met with during their day to day existences and their attempts to bring law and order to their mining camps. **8.5″ X 11″, 88 ppgs. Retail Price: $9.99**

The Geology and Mines of Northern Idaho and North Western Montana - Unavailable since 1909, this important publication was originally published by the Us Department of the Interior and has been unavailable for a century. Included are fine details on the geology and geography of the mining regions of Northern Idaho and North Western Montana. Of particular interest is a rare look at the mines and prospects of the region, including those in the Pine Creek Mining District, Lake Pend Oreille district, Troy Mining District, Sylvanite District, Cabinet Mining District, Prospect Mining District and the Missoula Valley. Some of the mines featured include the Iron Mountain, Silver Butte, Snowshoe, Grouse Mountain Mine and others. **8.5" X 11", 142 ppgs. Retail Price: $12.99**

Utah Mining Books

Fluorite in Utah - Unavailable since 1954, this publication was originally compiled by the USGS, State of Utah and U.S. Atomic Energy Commission and details the mining of fluorspar, also known as fluorite in the State of Utah. Included are details on the geology and history of fluorspar (fluorite) mining in Utah, including details on where this unique gem mineral may be found in the State of Utah. **8.5" X 11", 60 ppgs. Retail Price: $8.99**

California Mining Books

The Tertiary Gravels of the Sierra Nevada of California - Mining historian Kerby Jackson introduces us to a classic mining work by Waldemar Lindgren in this important re-issue of The Tertiary Gravels of the Sierra Nevada of California. Unavailable since 1911, this publication includes details on the gold bearing ancient river channels of the famous Sierra Nevada region of California. **8.5" X 11", 282 ppgs. Retail Price: $19.99**

The Mother Lode Mining Region of California - Unavailable since 1900, this publication includes details on the gold mines of California's famous Mother Lode gold mining area. Included are details on the geology, history and important gold mines of the region, as well as insights into historic mining methods, mine timbering, mining machinery, mining bell signals and other details on how these mines operated. Also included are insights into the gold mines of the California Mother Lode that were in operation during the first sixty years of California's mining history. **8.5" X 11", 176 ppgs. Retail Price: $14.99**

Lode Gold of the Klamath Mountains of Northern California and South West Oregon - Unavailable since 1971, this publication was originally compiled by Preston E. Hotz and includes details on the lode mining districts of Oregon and California's Klamath Mountains. Included are details on the geology, history and important lode mines of the French Gulch, Deadwood, Whiskeytown, Shasta, Redding, Muletown, South Fork, Old Diggings, Dog Creek (Delta), Bully Choop (Indian Creek), Harrison Gulch, Hayfork, Minersville, Trinity Center, Canyon Creek, East Fork, New River, Denny, Liberty (Black Bear), Cecilville, Callahan, Yreka, Fort Jones and Happy Camp mining districts in California, as well as the Ashland, Rogue River, Applegate, Illinois River, Takilma, Greenback, Galice, Silver Peak, Myrtle Creek and Mule Creek districts of South Western Oregon. Also included are insights into the mineralization and other characteristics of this important mining region. **8.5" X 11", 100 ppgs. Retail Price: $10.99**

Mines and Mineral Resources of Shasta County, Siskiyou County, Trinity County: California - Unavailable since 1915, this publication was originally compiled by the California State Mining Bureau and includes details on the gold mines of this area of Northern California. Also included are insights into the mineralization and other characteristics of this important mining region, as well as the location of historic gold mines. **8.5" X 11", 204 ppgs. Retail Price: $19.99**

Geology of the Yreka Quadrangle, Siskiyou County, California - Unavailable since 1977, this publication was originally compiled by Preston E. Hotz and includes details on the geology of the Yreka Quadrangle of Siskiyou County, California. Also included are insights into the mineralization and other characteristics of this important mining region. **8.5" X 11", 78 ppgs. Retail Price: $7.99**

Mines of San Diego and Imperial Counties, California - Originally published in 1914, this important publication on California Mining has not been available for a century. This publication includes important information on the early gold mines of San Diego and Imperial County, which were some of the first gold fields mined in California by early Spanish and Mexican miners before the 49ers came on the scene. Included are not only details on early mining methods in the area, production statistics and geological information, but also the location of the early gold mines that helped make California "The Golden State". Also included are details on the mining of other minerals such as silver, lead, zinc, manganese, tungsten, vanadium, asbestos, barite, borax, cement, clay, dolomite, fluospar, gem stones, graphite, marble, salines, petroleum, strontium, talc and others. **8.5" X 11", 116 ppgs. Retail Price: $12.99**

Mines of Sierra County, California - Unavailable since 1920, this publication was originally compiled by the California State Mining Bureau and includes details on the gold mines of Sierra County, California. Also included are insights into the mineralization and other characteristics of this important mining region, as well as the location of historic gold mines. **8.5" X 11", 156 ppgs. Retail Price: $19.99**

Mines of Plumas County, California - Unavailable since 1918, this publication was originally compiled by the California State Mining Bureau and includes details on the gold mines of Plumas County, California. Also included are insights into the mineralization and other characteristics of this important mining region, as well as the location of historic gold mines. 8.5" X 11", 200 ppgs. Retail Price: $19.99

Mines of El Dorado, Placer, Sacramento and Yuba Counties, California - Originally published in 1917, this important publication on California Mining has not been available for nearly a century. This publication includes important information on the early gold mines of El Dorado County, Placer County, Sacramento County and Yuba County, which were some of the first gold fields mined by the Forty-Niners during the California Gold Rush. Included are not only details on early mining methods in the area, production statistics and geological information, but also the location of the early gold mines that helped make California "The Golden State". Also included are insights into the early mining of chrome, copper and other minerals in this important mining area. 8.5" X 11", 204 ppgs. Retail Price: $19.99

Mines of Los Angeles, Orange and Riverside Counties, California - Originally published in 1917, this important publication on California Mining has not been available for nearly a century. This publication includes important information on the early gold mines of Los Angeles County, Orange County and Riverside County, which were some of the first gold fields mined in California by early Spanish and Mexican miners before the 49ers came on the scene. Included are not only details on early mining methods in the area, production statistics and geological information, but also the location of the early gold mines that helped make California "The Golden State". 8.5" X 11", 146 ppgs. Retail Price: $12.99

Mines of San Bernadino and Tulare Counties, California - Originally published in 1917, this important publication on California Mining has not been available for nearly a century. This publication includes important information on the early gold mines of San Bernadino and Tulare County, which were some of the first gold fields mined in California by early Spanish and Mexican miners before the 49ers came on the scene. Included are not only details on early mining methods in the area, production statistics and geological information, but also the location of the early gold mines that helped make California "The Golden State". Also included are details on the mining of other minerals such as copper, iron, lead, zinc, manganese, tungsten, vanadium, asbestos, barite, borax, cement, clay, dolomite, fluospar, gem stones, graphite, marble, salines, petroleum, stronium, talc and others. 8.5" X 11", 200 ppgs. Retail Price: $19.99

Chromite Mining in The Klamath Mountains of California and Oregon - Unavailable since 1919, this publication was originally compiled by J.S. Diller of the United States Department of Geological Survey and includes details on the chromite mines of this area of Northern California and Southern Oregon. Also included are insights into the mineralization and other characteristics of this important mining region, as well as the location of historic mines. Also included are insights into chromite mining in Eastern Oregon and Montana. 8.5" X 11", 98 ppgs. Retail Price: $9.99

Mines and Mining in Amador, Calaveras and Tuolumne Counties, California - Unavailable since 1915, this publication was originally compiled by William Tucker and includes details on the mines and mineral resources of this important California mining area. Included are details on the geology, history and important gold mines of the region, as well as insights into other local mineral resources such as asbestos, clay, copper, talc, limestone and others. Also included are insights into the mineralization and other characteristics of this important portion of California's Mother Lode mining region. 8.5" X 11", 198 ppgs. Retail Price: $14.99

The Cerro Gordo Mining District of Inyo County California - Unavailable since 1963, this publication was originally compiled by the United States Department of Interior. Included are insights into the mineralization and other characteristics of this important mining region of Southern California. Topics include the mining of gold and silver in this important mining district in Inyo County, California, including details on the history, production and locations of the Cerro Gordo Mine, the Morning Star Mine, Estelle Tunnel, Charles Lease Tunnel, Ignacio, Hart, Crosscut Tunnel, Sunset, Upper Newtown, Newtown, Ella, Perseverance, Newsboy, Belmont and other silver and gold mines in the Cerro Gordo Mining District. This volume also includes important insights into the fossil record, geologic formations, faults and other aspects of economic geology in this California mining district. 8.5" X 11", 104 ppgs. Retail Price: $10.99

Mining in Butte, Lassen, Modoc, Sutter and Tehama Counties of California - Unavailable since 1917, this publication was originally compiled by the United States Department of Interior. Included are insights into the mineralization and other characteristics of this important mining region of California. Topics include the mining of asbestos, chromite, gold, diamonds and manganese in Butte County, the mining of gold and copper in the Hayden Hill and Diamond Mountain mining districts of Lassen County, the mining of coal, salt, copper and gold in the High Grade and Winters mining districts of Modoc County, gold mining in Sutter County and the mining of gold, chromite, manganese and copper in Tehama County. This volume also includes the production records and locations of numerous mines in this important mining region. 8.5" X 11", 114 ppgs. Retail Price: $11.99

Mines of Trinity County California - Originally published in 1965, this important publication on California Mining has not been available for nearly fifty years. This publication includes important information on mines and mining in Trinity County, California, as well insights into the mineralization and geology of this important mining area in Northern California. Included are extensive details on hardrock and placer gold mines and prospects, including charts showing the locations of these historic mines.. 8.5″ X 11″, 144 ppgs. Retail Price: $12.99

Mines of Kern County California - Originally published in 1962, this important publication on California Mining has not been available for nearly fifty years. This publication includes important information on mines and mining in Kern County, California, as well insights into the mineralization and geology of this important mining area in California. Included are extensive details on hardrock and placer gold mines and prospects, including charts showing the locations of these historic mines. 8.5″ X 11″, 398 ppgs. Retail Price: $24.99

Mines of Calaveras County California - Originally published in 1962, this important publication on California Mining has not been available for nearly fifty years. This publication includes important information on mines and mining in Calaveras County, California, as well insights into the mineralization and geology of this important mining area in Northern California. Included are extensive details on hardrock and placer gold mines and prospects, including charts showing the locations of these historic mines. 8.5″ X 11″, 236 ppgs. Retail Price: $19.99

Lode Gold Mining in Grass Valley California - Unavailable since 1940, this publication was originally compiled by the United States Department of Interior. Included are insights into the gold mineralization and other characteristics of this important mining region of Nevada County, California. This volume also includes important insights into the geologic formations, faults and other aspects of economic geology in this California mining district. Of particular interest are the fine details on many hardrock gold mines in the area, including their locations, histories, development and mineralization. Some of the mines featured include the Gold Hill Mine, Massachusetts Hill, Boundary, Peabody, Golden Center, North Star, Omaha, Lone Jack, Homeward Bound, Hartery, Wisconsin, Allison Ranch, Phoenix, Kate Hayes, W.Y.O.D., Empire, Rich Hill, Daisy Hill, Orleans, Sultana, Centennial, Conlin, Ben Franklin, Crown Point and many others. 8.5″ X 11″, 148 ppgs. Retail Price: $12.99

Alaska Mining Books

Ore Deposits of the Willow Creek Mining District, Alaska - Unavailable since 1954, this hard to find publication includes valuable insights into the Willow Creek Mining District near Hatcher Pass in Alaska. The publication includes insights into the history, geology and locations of the well known mines in the area, including the Gold Cord, Independence, Fern, Mabel, Lonesome, Snowbird, Schroff-O'Neil, High Grade, Marion Twin, Thorpe, Webfoot, Kelly-Willow, Lane, Holland and others. 8.5″ X 11″, 96 ppgs. Retail Price: $9.99

Arizona Mining Books

Mines and Mining in Northern Yuma County Arizona - Originally published in 1911, this important publication on Arizona Mining has not been available for over a hundred years. Included are rare insights into the gold, silver, copper and quicksilver mines of Yuma County, Arizona together with hard to find maps and photographs. Some of the mines and mining districts featured include the Planet Copper Mine, Mineral Hill, the Clara Consolidated Mine, Viati Mine, Copper Basin prospect, Bowman Mine, Quartz King, Billy Mack, Carnation, the Wardwell and Osbourne, Valensuella Copper, the Mariquita, Colonial Mine, the French American, the New York-Plomosa, Guadalupe, Lead Camp, Mudersbach Copper Camp, Yellow Bird, the Arizona Northern (Salome Strike), Bonanza (Harqua Hala), Golden Eagle, Hercules, Socorro and others. 8.5″ X 11″, 144 ppgs. Retail Price: $11.99

The Aravaipa and Stanley Mining Districts of Graham County Arizona - Originally published in 1925, this important publication on Arizona Mining has not been available for nearly ninety years. Included are rare insights into the gold and silver mines of these two important mining districts, together with hard to find maps. 8.5″ X 11″, 140 ppgs. Retail Price: $11.99

Gold in the Gold Basin and Lost Basin Mining Districts of Mohave County, Arizona - This volume contains rare insights into the geology and gold mineralization of the Gold Basin and Lost Basin Mining Districts of Mohave County, Arizona that will be of benefit to miners and prospectors. Also included is a significant body of information on the gold mines and prospects of this portion of Arizona. This volume is lavishly illustrated with rare photos and mining maps. 8.5″ X 11″, 188 ppgs. Retail Price: $19.99

Mines of the Jerome and Bradshaw Mountains of Arizona - This important publication on Arizona Mining has not been available for ninety years. This volume contains rare insights into the geology and ore deposits of the Jerome and Bradshaw Mountains of Arizona that will be of benefit to miners and prospectors who work those areas. Included is a significant body of information on the mines and prospects of the Verde, Black Hills, Cherry Creek, Prescott, Walker, Groom Creek, Hassayampa, Bigbug, Turkey Creek, Agua Fria, Black Canyon, Peck, Tiger, Pine Grove, Bradshaw, Tintop, Humbug and Castle Creek Mining Districts. This volume is lavishly illustrated with rare photos and mining maps. 8.5″ X 11″, 218 ppgs. Retail Price: $19.99

The Ajo Mining District of Pima County Arizona - This important publication on Arizona Mining has not been available for nearly seventy years. This volume contains rare insights into the geology and mineralization of the Ajo Mining District in Pima County, Arizona and in particular the famous New Cornelia Mine.   8.5" X 11", 126 ppgs. Retail Price: $11.99

Mining in the Santa Rita and Patagonia Mountains of Arizona - Originally published in 1915, this important publication on Arizona Mining has not been available for nearly a century. Included are rare insights into hundreds of gold, silver, copper and other mines in this famous Arizona mining area. Details include the locations, geology, history, production and other facts of the mines of this region. 8.5" X 11", 394 ppgs. Retail Price: $24.99

Montana Mining Books

A History of Butte Montana: The World's Greatest Mining Camp - First published in 1900 by H.C. Freeman, this important publication sheds a bright light on one of the most important mining areas in the history of The West. Together with his insights, as well as rare photographs of the periods, Harry Freeman describes Butte and its vicinity from its early beginnings, right up to its flush years when copper flowed from its mines like a river. At the time of publication, Butte, Montana was known worldwide as "The Richest Mining Spot On Earth" and produced not only vast amounts of copper, but also silver, gold and other metals from its mines. Freeman illustrates, with great detail, the most important mines in the vicinity of Butte, providing rare details on their owners, their history and most importantly, how the mines operated and how their treasures were extracted. Of particular interest are the dozens of rare photographs that depict mines such as the famous Anaconda, the Silver Bow, the Smoke House, Moose, Paulin, Buffalo, Little Minah, the Mountain Consolidated, West Greyrock, Cora, the Green Mountain, Diamond, Bell, Parnell, the Neversweat, Nipper, Original and many others.   8.5" X 11", 142 ppgs. Retail Price: $12.99

The Butte Mining District of Montana - This important publication on Montana Mining has not been available for over a century. Included are rare insights into the gold, copper and silver mines of Butte, Montana together with hard to find maps and photographs. Some of the topics include the early history of gold, silver and copper mining in the Butte area, insight into the geology of its mining areas, the local distribution of gold, silver and copper ores, as well their composition and how to identify them. Also included are detailed facts about the mines in the Butte Mining District, including the famous Anaconda Mine, Gagnon, Parrot, Blue Vein, Moscow, Poulin, Stella, Buffalo, Green Mountain, Wake Up Jim, the Diamond-Bell Group, Mountain Consolidated, East Greyrock, West Greyrock, Snowball, Corra, Speculator, Adirondack, Miners Union, the Jessie-Edith May Group, Otisco, Iduna, Colorado, Lizzie, Cambers, Anderson, Hesperus, Preferencia and dozens of others. 8.5" X 11", 298 ppgs. Retail Price: $24.99

Mines of the Helena Mining Region of Montana - This important publication on Montana Mining has not been available for over a century. Included are rare insights into the gold, copper and silver mines of the vicinity of Helena, Montana, including the Marysville Mining District, Elliston Mining District, Rimini Mining District, Helena Mining District, Clancy Mining District, Wickes Mining District, Boulder and Basin Mining Districts and the Elkhorn Mining District. Some of the topics include the early history of gold, silver and copper mining in the Helena area, insight into the geology of its mining areas, the local distribution of gold, silver and copper ores, as well their composition and how to identify them. Also included are detailed facts, history, geology and locations of over one hundred gold, silver and copper mines in the area . 8.5" X 11", 162 ppgs, Retail Price: $14.99

Mines and Geology of the Garnet Range of Montana - This important publication on Montana Mining has not been available for over a century. Included are rare insights into the gold, copper and silver mines of the vicinity of this important mining area of Montana. Some of the topics include the early history of gold, silver and copper mining in the Garnet Mountains, insight into the geology of its mining areas, the local distribution of gold, silver and copper ores, as well their composition and how to identify them. Also included are detailed facts, history, geology and locations of numerous gold, silver and copper mines in the area . 8.5" X 11", 100 ppgs, Retail Price: $11.99

Mines and Geology of the Philipsburg Quadrangle of Montana - This important publication on Montana Mining has not been available for over a century. Included are rare insights into the gold, copper and silver mines of the vicinity of this important mining area of Montana. Some of the topics include the early history of gold, silver and copper mining in the Philipsburg Quadrangle, insight into the geology of its mining areas, the local distribution of gold, silver and copper ores, as well their composition and how to identify them. Also included are detailed facts, history, geology and locations of over one hundred gold, silver and copper mines in the area 8.5" X 11", 290 ppgs, Retail Price: $24.99

Geology of the Marysville Mining District of Montana - Included are rare insights into the mining geology of the Marysville Mining District. Some of the topics include the early history of gold, silver and copper mining in the area, insight into the geology of its mining areas, the local distribution of gold, silver and copper ores, as well their composition and how to identify them. Also included are detailed facts, history, geology and locations of gold, silver and copper mines in the area 8.5" X 11", 198 ppgs, Retail Price: $19.99

<u>The Geology and Mines of Northern Idaho and North Western Montana</u>

See listing under Idaho.

Nevada Mining Books

<u>The Bull Frog Mining District of Nevada</u> - Unavailable since 1910, this publication was originally compiled by the United States Department of Interior. This volume also includes important insights into the geologic formations, faults and other aspects of economic geology in this Nevada mining district. Of particular interest are the fine details on many mines in the area, including their locations, histories, development and mineralization. Some of the mines featured include the National Bank Mine, Providence, Gibraltor, Tramps, Denver, Original Bullfrog, Gold Bar, Mayflower, Homestake-King and other mines and prospects. **8.5" X 11", 152 ppgs, Retail Price: $14.99**

Colorado Mining Books

<u>Ores of The Leadville Mining District</u> - Unavailable since 1926, this publication was originally compiled by the United States Department of Interior. This volume also includes important insights into the ores and mineralization of the Leadville Mining District in Colorado. Topics include historic ore prospecting methods, local geology, insights into ore veins and stockworks, the local trend and distribution of ore channels, reverse faults, shattered rock above replacement ore bodies, mineral enrichment in oxidized and sulphide zones and more. **8.5" X 11", 66 ppgs, Retail Price: $8.99**

<u>Mining in Colorado</u> - Unavailable since 1926, this publication was originally compiled by the United States Department of Interior. This volume also includes important insights into the mining history of Colorado from its early beginnings in the 1850's right up to the mid 1920's. Not only is Colorado's gold mining heritage included, but also its silver, copper, lead and zinc mining industry. Each mining area is treated separately, detailing the development of Colorado's mines on a county by county basis. **8.5" X 11", 284 ppgs, Retail Price: $19.99**

<u>Gold Mining in Gilpin County Colorado</u> - Unavailable since 1876, this publication was originally compiled by the Register Steam Printing House of Central City, Colorado. A rare glimpse at the gold mining history and early mines of Gilpin County, Colorado from their first discovery in the 1850's up to the "flush years" of the mid 1870's. Of particular interest is the history of the discovery of gold in Gilpin County and details about the men who made those first strikes. Special focus is given to the early gold mines and first mining districts of the area, many of which are not detailed in other books on Colorado's gold mining history. **8.5" X 11", 156 ppgs, Retail Price: $12.99**

<u>Mining in the Gold Brick Mining District of Colorado</u> - Important insights into the history of the Gold Brick Mining District, as well as its local geography and economic geology. Also included are the histories and locations of historic mines in this important Colorado Mining District, including the Cortland, Carter, Raymond, Gold Links, Sacramento, Bassick, Sandy Hook, Chronicle, Grand Prize, Chloride, Granite Mountain, Lucille, Gray Mountain, Hilltop, Maggie Mitchell, Silver Islet, Revenue, Roosevelt, Carbonate King and others. In addition to hardrock mining, are also included are details on gold placer mining in this portion of Colorado. **8.5" X 11", 140 ppgs, Retail Price: $12.99**

Washington Mining Books

<u>The Republic Mining District of Washington</u> - Unavailable since 1910, this important publication was originally published by the Washington Geologic Survey and has been unavailable for a century. Topics include the geology, rock formations and the formation of ore deposits in this important mining area of Washington State. Also included are hard to find details on the geology, history and locations of dozens of mines in the area. Some of the mines featured include the New Republic Mine, Ben Hur, Morning Glory, the South Republic Mine, Quilp, Surprise, Black Tail, Lone Pine, San Poil, Mountain Lion, Tom Thumb, Elcaliph and many others. **8.5" X 11", 94 ppgs, Retail Price: $10.99**

Wyoming Mining Books

<u>Mining in the Laramie Basin of Wyoming</u> - Unavailable since 1909, this publication was originally compiled by the United States Department of Interior. Also included are insights into the mineralization and other characteristics of this important mining region, especially in regards to coal, limestone, gypsum, bentonite clay, cement, sand, clay and copper. **8.5" X 11", 104 ppgs, Retail Price: $11.99**

More Mining Books

Prospecting and Developing A Small Mine - Topics covered include the classification of varying ores, how to take a proper ore sample, the proper reduction of ore samples, alluvial sampling, how to understand geology as it is applied to prospecting and mining, prospecting procedures, methods of ore treatment, the application of drilling and blasting in a small mine and other topics that the small scale miner will find of benefit. **8.5" X 11", 112 ppgs, Retail Price: $11.99**

Timbering For Small Underground Mines - Topics covered include the selection of caps and posts, the treatment of mine timbers, how to install mine timbers, repairing damaged timbers, use of drift supports, headboards, squeeze sets, ore chute construction, mine cribbing, square set timbering methods, the use of steel and concrete sets and other topics that the small underground miner will find of benefit. This volume also includes twenty eight illustrations depicting the proper construction of mine timbering and support systems that greatly enhance the practical usability of the information contained in this small book. **8.5" X 11", 88 ppgs. Retail Price: $10.99**

Timbering and Mining - A classic mining publication on Hard Rock Mining by W.H. Storms. Unavailable since 1909, this rare publication provides an in depth look at American methods of underground mine timbering and mining methods. Topics include the selection and preservation of mine timbers, drifting and drift sets, driving in running ground, structural steel in mine workings, timbering drifts in gravel mines, timbering methods for driving shafts, positioning drill holes in shafts, timbering stations at shafts, drainage, mining large ore bodies by means of open cuts or by the "Glory Hole" system, stoping out ore in flat or low lying veins, use of the "Caving System", stoping in swelling ground, how to stope out large ore bodies, Square Set timbering on the Comstock and its modifications by California miners, the construction of ore chutes, stoping ore bodies by use of the "Block System", how to work dangerous ground, information on the "Delprat System" of stoping without mine timbers, construction and use of headframes and much more. This volume provides a reference into not only practical methods of mining and timbering that may be employed in narrow vein mining by small miners today, but also rare insights into how mines were being worked at the turn of the 19th Century. **8.5" X 11", 288 ppgs. Retail Price: $24.99**

A Study of Ore Deposits For The Practical Miner - Mining historian Kerby Jackson introduces us to a classic mining publication on ore deposits by J.P. Wallace. First published in 1908, it has been unavailable for over a century. Included are important insights into the properties of minerals and their identification, on the occurrence and origin of gold, on gold alloys, insights into gold bearing sulfides such as pyrites and arsenopyrites, on gold bearing vanadium, gold and silver tellurides, lead and mercury tellurides, on silver ores, platinum and iridium, mercury ores, copper ores, lead ores, zinc ores, iron ores, chromium ores, manganese ores, nickel ores, tin ores, tungsten ores and others. Also included are facts regarding rock forming minerals, their composition and occurrences, on igneous, sedimentary, metamorphic and intrusive rocks, as well as how they are geologically disturbed by dikes, flows and faults, as well as the effects of these geologic actions and why they are important to the miner. Written specifically with the common miner and prospector in mind, the book will help to unlock the earth's hidden wealth for you and is written in a simple and concise language that anyone can understand. **8.5" X 11", 366 ppgs. Retail Price: $24.99**

Mine Drainage - Unavailable since 1896, this rare publication provides an in depth look at American methods of underground mine drainage and mining pump systems. This volume provides a reference into not only practical methods of mining drainage that may be employed in narrow vein mining by small miners today, but also rare insights into how mines were being worked at the turn of the 19th Century. **8.5" X 11", 218 ppgs. Retail Price: $24.99**

Fire Assaying Gold, Silver and Lead Ores - Unavailable since 1907, this important publication was originally published by the Mining and Scientific Press and was designed to introduce miners and prospectors of gold, silver and lead to the art of fire assaying. Topics include the fire assaying of ores and products containing gold, silver and lead; the sampling and preparation of ore for an assay; care of the assay office, assay furnaces; crucibles and scorifiers; assay balances; metallic ores; scorification assays; cupelling; parting' crucible assays, the roasting of ores and more. This classic provides a time honored method of assaying put forward in a clear, concise and easy to understand language that will make it a benefit to even beginners. **8.5" X 11", 96 ppgs. Retail Price: $11.99**

Methods of Mine Timbering - Originally published in 1896, this important publication on mining engineering has not been available for nearly a century. Included are rare insights into historical methods of timbering structural support that were used in underground metal mines during the California that still have a practical application for the small scale hardrock miner of today. **8.5" X 11", 94 ppgs. Retail Price: $10.99**